FARMS OF TO

Community Supported Farms
Farm Supported Communities

by
Trauger Groh and
Steven S.H. McFadden

© 1990

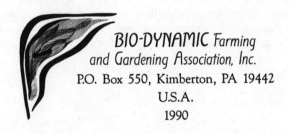

BIO-DYNAMIC Farming
and Gardening Association, Inc.
P.O. Box 550, Kimberton, PA 19442
U.S.A.
1990

© Bio-Dynamic Literature 1990

ISBN 0-938250-28-0

Printed in U.S.A.

Typesetting by MARILYN McDONALD-FRANCIS
Cover by LEAH KRISTIN COTTER

 PRINTED ON 80%
RECYCLED FIBER

Table of Contents

Part III

Acknowledgements

This author has had the good fortune to be included in the serious striving for the farms of tomorrow taking place in both the old and the new world. If not for this good fortune, this book would not have been written, and some of the farms discussed here could not have been established. Some of the farmers involved in this striving I am proud to call my friends. They have struggled by themselves and in cooperation with others to develop new concepts. They have been ready to take great risks trying new ways with their land, with their labor, and in addressing the basic questions of farming that are described in this book. Some of these farmers were very much alone in this striving at first, but they have held onto their ideals to develop a greater understanding for new methods.

Two close friends of this author were particularly helpful: Heiloh Loss and Carl-August Loss. They decided in 1968 to donate the only property they had, a 200-acre farm that carried practically no mortgage, to a land trust. This action made possible an innovative experiment in cooperation with the author and with other people on this free land. The new farm they created together, Buschberg-Hof, has proved that free individuals can cooperate in a farm operation that works land held in trust, rather than as private property. This experiment also established the possibility that a farm can succeed when it strives for quality rather than profit, and that by this the farmers can support themselves and their families more securely than if they had been running the farm for profit.

In addition to the farmers who helped this author put new ideas into practice, two persons have to be mentioned who gave him impulses and enthusiasm out of their life experience in the field of

1

agriculture and social economics. Dr. Nicolaus Remer, a farmer and agriculturist who worked as a researcher and farm adviser with hundreds of farmers over 50 years, allowed the author to take part in his spiritual striving as an agriculturist, and shared his vast experience. He gave his help in many ways in establishing ecologically sound farm organisms based on a new social concept. Dr. Remer belongs to the group of farmers and scientists who developed the biodynamic approach to agriculture through practical application and continuous research. As a result of these pioneering efforts, we can say today: biodynamics rightly applied can provide mankind with ample healthy lifegiving food, and can not only conserve our environment but improve it where it has been neglected or abused.

This author is also deeply indebted to Wilhelm Ernst Barkhoff, a lawyer and economist who has taken a deep interest in the socio-economic problems of farming. He gave me and many other farmers the inspiration to rethink the legal, social, and economic conditions of agriculture, and helped us to set up models where people could cooperate agriculturally on free land, including even non-farmers who took an interest in this task. I learned from him to leave worn-out tracks of thinking and start a fresh approach in an unprepossessed way. Many years ago he gave me three golden rules for community farming that have proved precious and valid for any such endeavor. I am pleased to present those rules later in this book.

Finally, I want to express my gratitude to Alice Bennett-Groh. She carried on all the consequences of family life that arise out of the fact that I am away from home learning, lecturing and advising on these matters, besides writing manuscripts and participating actively in the Temple-Wilton Community Farm. Beyond that she is ready to discuss every thought and every measure that goes with this activity in an understanding and helpful way.

It is just not possible to acknowledge the hundreds of books and articles that I have read on farming and related matters. Yet I realize and acknowledge that many sources have influenced and helped me to shape the picture I have tried to present in this book. For all of those writers and their many insights, I am deeply grateful.

— Trauger Groh

My earliest interest in the garden arose from observations of my mother's efforts with a bed of reluctant peonies. Though I was also reluctant, she drafted me as a child laborer. Somehow, through sharing with her an intense and oftentimes frustrating interest in that bed of flowers, and also the love of mother and child, within me was planted the seed that in later life blossomed into a consuming interest in gardening and farming.

My primary teachers have been the plant and mineral kingdoms themselves, especially the flowers with their many soul-healing attributes. And so to these realms I give deep thanks.

I also wish to acknowledge my neighbors, the farmers of the Temple-Wilton Community Farm: Lincoln Geiger, Anthony Graham, and Trauger Groh. Their deep thought, their beautifully coordinated efforts, and the impressive results of their work, have steadily inspired me to advance my own cultivation practices. Much more lies ahead.

— Steven S.H. McFadden

Introduction

"When tillage begins, other arts follow. The farmers therefore are the founders of civilization."

> — Daniel Webster
> New Hampshire statesman

Agriculture is the foundation of modern civilization. Without a steady supply of clean, life-giving food, we have neither the leisure nor the energy to develop industry, science or art. Worldwide and in particular in the United States, our foundation has deteriorated dangerously. It requires immediate and fundamental restructuring. But how can we even begin to approach this task?

This book has been written to suggest some possibilities, and also to serve a need that is becoming more and more explicit: the need to share the experience of farming with everyone who understands that our relationship with nature and the ways that we use the land will determine the future of the earth. The problems of agriculture and the environment belong not just to a small minority of active farmers; they are the problems of all humanity, and thousands of people are searching for new ways and new solutions.

As the farming crisis deepens, many people are seeking wiser, more effective ways to reestablish the relationship of human beings with the earth. The financial and agricultural practices of recent decades have made it increasingly difficult, and in some cases impossible, for existing models of agriculture to prosper. In America, the family farm has fallen victim to a relentless marketplace; meanwhile, corporate farms have tended to place short-run economic advantage over the long-term considerations of our relationship with each other and the earth. Modern ways of industrial and chemical farming play a major part in the deterioration of our environment on all levels: soil, water, air, landscape, and plant and animal life. Only a new, ecologically sound approach to farming can slow down or stop this deterioration.

When we look to the universal questions of land use and land abuse, and see the manifold dimensions of these questions, we understand quickly that there is no universal solution. There is no simple recipe or remedy for the many challenges we face. Out of this understanding, the authors decided upon the following approach: first, to work out some fundamental questions and principles of land use as it concerns our food, our environment, and our general ways of living with the land. Second, to present living examples of a new approach to the use of land. And third, to offer readers a list of resources so that they may have ready access to information which will support them in the pursuit of new, healthier uses of the land.

Part I was written by Trauger Groh, and it represents the fruit of 30 years of experience in practical farming and advisory work, as well as numerous lectures in various countries of both the new and the old world. Part II, the descriptive part, was written by Steven S.H. McFadden, a journalist with a special interest in ecological and agricultural questions. The resources at the end of the book were gathered by Steven and by Rod Shouldice of the Bio-Dynamic Association.

The experiments in farming described in this book represent new social forms of agriculture which have arisen in recent years while traditional family farms have declined and industrial agriculture has increased. These farms are not static organizations, but rather living organisms which constantly change. Many of them have, in fact, changed considerably during the writing and publication of this book. These new farms involve many local families directly in the decisions and labor which produce the vegetables, fruits, milk, and meat they eat. In that way they reestablish a link between the farm, the farmer, and the consumer. While this approach may not be the full answer to the questions posed by the modern agricultural dilemma, we believe it has much to offer.

The authors tried to select examples that show great variety in approaches to the farms of tomorrow. Generally these approaches are run in America under the name "Community Supported Agriculture" (CSA). As with many catchall names, the term community supported agriculture or CSA is slightly misleading. It implies that the problem is special support for agriculture. As important and necessary as that may be, it is secondary. Although it may seem a fine point, the primary need is not for the farm to be supported by the community, but rather for the community to support itself through farming. This is an essential of existence, not a matter of convenience. We have no choice about whether to farm or not, as we have a choice about whether to produce TV sets or not. So we have to either farm or to support farmers, every one of us, at any cost. We cannot give it up because it is inconvenient or unprofitable.

Since our existence is primarily dependent on farming, we cannot entrust this essential activity solely to the farming population — just 2% of Americans. As farming becomes more and more remote from the life of the average person, it becomes less and less able to provide us with clean, healthy, lifegiving food or a clean, healthy, lifegiving environment. A small minority of farmers, laden with debt and overburdened with responsibility, cannot possibly meet the needs of all the people.

More and more people are coming to recognize this, and they are becoming ready to share agricultural responsibilities with the active farmers. Out of this impulse, many CSAs have developed in America from 1985 to 1990. Out of these, the authors have selected seven different farms as models. In recognition that the deterioration of our farm system is frequently caused by our monetary system, we have made a special effort to explore new ways of farm financing.

Some things are typical for all community supported farms. In all of them there is a strong dedication to quality; most of them are organic or biodynamic farms, most of them show great diversification, most are integrated farm organisms having their own livestock and thus their own source of manure, or they are aiming in this direction. At all of them, far more people are working regularly per 100 acres than in conventionally run farms; and generally there are just many more people around participating in all the dimensions of agricultural life: working, relaxing, storing, shopping, celebrating. This human element is of enormous importance. It shows that these farms have something to offer beyond good food. They embody educational and cultural elements that draw the interest of many people. Besides clean, healthy, lifegiving food, and a strong contribution to an improved environment, the educational and cultural elements constitute the third great gift that the farms of tomorrow have to offer.

Neither the urban nor the suburban lifestyle of today are able to provide the fullness of experiences that the human being needs for its development. In the future, as in the past, everybody, especially in childhood and in youth, needs the soul and body nourishing experience that only the active and creative engagement with nature in gardens and life-filled farm organisms can give.

PART 1

Essays on the Farms of Tomorrow

by Trauger Groh

ESSAY 1
Why Do We Need New Farms?
Food — Environment — Education

When we speak about the need for healthy farm organisms, we think first of our food supply and then we think of the farm as part of our natural world, shaping the environment in positive or negative ways. Rarely do we have in mind the great contribution that living on farms and working in nature gives to our inner soul development and to the shaping of our social faculties. Yet all three of these considerations are essential elements of agriculture, and of the farms of tomorrow.

Healthy Food
The question of food and food quality is very complex. We speak in general terms about healthy food, or life-giving food. But these terms can mean different things to different people. In the modern context, perhaps a more accessible concept is "clean food," clean meaning free of any synthetic substances that might be added during growing, processing or preserving. Such substances are typically preservatives, insecticides, fungicides, herbicides, synthetic colors, and so forth. Arguments about which additives are tolerable and

8

which pose a health threat are complex and confusing to most people, and so we let the government step in to make such determinations. But we should be skeptical towards authorities who decide these questions for us. It is extremely time-consuming and difficult to establish the exact health effect of any of the many synthetic substances that are routinely added to our food. One thing we can say with certainty. The cumulative effect of the different substances that are added is largely unknown. Government agencies such as the Food and Drug Administration (FDA) are simply not in a position to guarantee the safety of any food additive, even if they pretend to be. They are not even able to test properly what is in use, never mind the new synthetic substances constantly being introduced to the market. The standard declaration of additives on food packaging is a good thing, but the widespread belief that what is declared, and therefore allowed by the government, is without problems is an illusion.

That leaves the wise consumer only one choice: to demand food without any additives. If we ask for a carrot, we should demand carrot, and only what nature gives us in the carrot. If we ask for milk we should demand milk in the beautiful composition given by a properly fed cow, not accepting anything more or less, such as the synthetic vitamins typically added during processing, or the loss of the life essence that occurs during pasturization.

Can we, or should we, allow the state to make the basic decisions about what is good for us? Is this not a basic right of the citizen? In the case of milk, for example, the government has assumed the right to decide that milk as nature gives it is hazardous to human health, and that therefore all milk must be heat processed in a way that changes markedly its natural composition, robbing it of essential parts and driving out all the life forces that are in it. If someone wants to consume raw milk or some other forbidden food, and if that person believes the food is good, and also has a trusting relationship with the farmer who produces the food, should they not have that right?

The absence or presence of additives alone does not determine the quality of food. The fundamental secret of quality production is to handle the plants and animals so that they attain their highest performance by their own nature. In each creation, there is an inner harmony of substances and forces that is typical and healthy. It is not the presence of certain substances in certain amounts that makes a vegetable or grain healthy; rather, it is the harmonious relationship between the substances and the forces. To a large extent, modern agricultural methods have drastically affected this harmony. As research has shown, already between 1896 and 1932 many crops exhibited a strong rise in the content of potash while their magnesium content declined. Meanwhile, other research shows that the silica content in cultivated plants has tended to decline while the potash content has been rising.*

The results of this change to a less harmonious balance showed up in Eastern Europe, where for hundreds of years people thatched their roofs with rye straw. Those roofs typically lasted for 15 years. But after the rye crops were treated with synthetic nitrogen, and the natural harmony of substances and forces had been altered, the roofs fashioned from the resulting straw began to rot after just three to five years. Though perhaps not so obvious, similar changes have occurred in the bread grain that is a staple of our diet. There the weakening of the plants through unharmonious fertilization shows up in the excessive appearance of fungus diseases, which again provokes the use of harsh fungicides. As for the grain itself, the potash and phosphorous content is higher today than 100 years ago, and the silica content is less. What influence does this profound change have on the human beings who eat the bread and other products made from this grain? Some observers believe the high phosphorous content in many processed foods, much of which comes through industrial food processing, is a major factor in problems of hyperactive children, and other observers believe that the reduced silica content has led to a dulling of our senses.

While science has developed highly sophisticated ways of making quantitative measurements, the concept of quality is difficult to measure with gauges and scales. To evaluate quality, we must observe how the food affects the higher organisms who digest it. For example, carefully designed tests have conclusively demonstrated the effect of organic, biodynamic, and conventionally grown grains

---

* *Lebensgesetze im Landbau,* Remer (1968) *The Dynamics of Nutrition,* Schmidt, 1980.

upon the urine of domestic animals. If the quality of the food can be detected in the excretion of an organism, then clearly the quality of the food is also having an effect on the health of the organism.

As we create the farms and the culture of tomorrow, we need to aim in a certain direction with our nutrition. What do we want to achieve with nutrition besides keeping up our bodily functions? How can our diet support not only our physical health but also the development of our spiritual faculties so that they function in the best way? The point that men and women live longer today than in the past is a poor argument for the quality of our food if we do not pose questions about the condition of our lives. What do we want to achieve in life? Are we really in full possession of our faculties of thinking, feeling, and willing? One striking example of dulled spiritual faculties comes in the realm of free will. In general, modern men and women have strong will forces, but their lives lack direction and creativity; the will forces are not channeled in the service of creativity. Common deficiencies in enacting ones will forces and the moral insanity that we perceive all around us may well be connected to the low quality of the food generally available for consumption.

Food quality is first determined upon the farm by the way we interact with nature and its forces. The profit motivation does not lead to quality of food production. This thesis can be proved by looking into the history of modern farming in the last 100 years and into the state of affairs with our processed foods. Farming differs here from the production and marketing of industrial goods. You cannot sell cars that have grave deficiencies for very long, but you can deceive mankind for a long time with deficient food. The consequences of a deficient car show up very rapidly, but the effects of deficient food — nicely colored and flavored with artificial ingredients — are much harder to discern, and turn up mainly in the soul life of humanity or in the health problems of old age.

Nearly all manipulations with food — additives, radiation, and conservation methods — serve not the purpose of quality, but rather the purpose of distribution over long distances, shelf-life, and a pleasing appearance. Contrary to what might be right for many industrial products, the production, processing, distribution, and consumption of food favors quality when it is done locally. At the same time, this is the most economic approach to food because it saves transportation and preservation costs. The community supported farm systems of the future will proceed in this way; that is, producing for the local community, which includes the closest cities. Here households will connect themselves with local farms direcly or via trusted agents so that they can support a system of production that aims primarily at quality rather than profit.

A Healthy Environment

Protection of the earth, the air, and the water, and the positive development of an already desecrated environment is not possible without healthy farm organisms. This becomes obvious, for example, in those parts of New England where farming has ceased to exist, and where the landscape consists only of overgrown farmland and suburban development. Once the farms with their animals are gone, and with them the open fields and pastures, the scenery loses its soul-animating quality. On the other hand, we can see contemporary industrial farms as one of the great destroyers of the environment. The vast groundwater pollution of our days, the extreme erosion of topsoil, these are just some of the consequences of modern, profit-oriented farm systems. As they are established, the farms of tomorrow must protect against the following abuses:

• The soil against one-sided agricultural uses, and the poisonous effects of artificial fertilizer, pesticides, herbicides and fungicides.

• The water against pollution, lowering of the water table, and the consequences of soil erosion.

• The air against poisonous emissions.

• The life of wild plants and wild animals against the dwindling of natural habitats, leading to eventual extinction.

• The cultivated plants and domesticated animals against degeneration and exploitative management practices, including biological engineering.

• The landscape against monotony and sterility.

All this we can achieve only with the help of ecologically sound farm organisms that, while producing the necessary food for humanity, not only protect the environment but also cultivate the land in a healthy way.

Ecologically sound farm organisms mean a new type which there is created and upheld a balanced relationship bec. animal husbandry, field and pasture on one hand, and forest, hedgerow, water, and fallow land on the other.

If we envisage this type of farm — and we have some good examples among the community supported farms described in this book — and we compare them with the monotony of modern farm industry, then we become aware what the farms of the future will mean for our well being and for the future of the planet earth.

The Education of Humanity

Besides quality food production and environmental care, there is a third essential task which the farms of today in their one-sided industrial production have tended to neglect, and which the farms of the future must take up again. That task is the education of mankind through active work in nature, specifically nature that is formed into healthy, self-supporting, ecologically sound farm organisms. If we look back into the agricultural societies of the past, and this past lies only 200 to 300 years back, we find a very simple school system. For those children who went to school at all, it was thought sufficient if they learned the stories of the holy books, to read and to write, and perhaps some mathematics. But for most people, the great educators of olden times were the farm and nature. The farms of tomorrow can also teach many profound lessons, including the following:

• To lead a rhythmic lifestyle that is shaped by the seasons and the rhythms of the day — that is, the rhythms of the sun. In the modern world we have liberated ourselves from natural rhythms through electric lights and the use of machines. However, people who use the artificial light to completely reverse day and night tend to become ill. They are working against the rhythm of nature, rather than with it. The rhythms of nature are shaped by the rhythms of the larger cosmos and also live in our body. When we observe these rhythms and bring our daily life into harmonious accord with them, we unite healthy forces with our body and our soul life.

• A modest lifestyle adapted to what nature locally, with the help of a farmer, can give. When we are in touch with the capacities and limitations of nature, this awareness spills over into other dimensions of life and helps establish balance and modesty. In the

13

industrialized world this modesty has been lost. There is a feeling
that everything is possible if there is enough money. The result of
this attitude is a nation that consumes 24% of the world's energy
and produces 30% of the world's garbage, even though it contains
only 5% of the world's population. It is choking on its own waste
as the landfills overflow. The farms of tomorrow can naturally
instruct humanity in a lifestyle that is not only more modest, but
also more satisfying.

• A readiness to do what is necessary without complaint —
feeding and cleaning the animals, caring for and harvesting the crops,
and sublimating one's pleasures to these duties. Unlike the demands of
parents or teachers, the demands of the farm can be easier understood
and accepted by young people. When a parent demands that a child
behave, or do homework, those demands can seem abstract. But when
corn must be harvested, or sheep fed, the demands are tangible, and
this teaches the value of work and of service.

• A deep feeling for the self-evident fact that you cannot harvest
anything without planting. This is fundamental logic, in concrete rather
than abstract terms.

• An appreciation that in reality it is nature that produces on the
farm, not the farmer, and that natural production on the farm is not
an input-output equation, but rather a cooperative venture with the
forces of the earth and the cosmos. The only new wealth in the world
comes from the forests and the fields, as each year nature renews itself.
As it stands now, farming is a wasteful enterprise, for the input of
substance and energy is often higher than the output. This is inherently
unbalanced and is leading to many deep problems. Such an
understanding has enormous importance for our economic world view.

• An understanding that working on farms means relying on what
other generations have done — clearing the land, draining the
swamp, picking the stones out, and so forth. From this understanding
arises the recognition that your own contribution will not necessarily
serve you and your family alone, but also future generations. If you
plant a white oak or a redwood tree, the time of its maturity will
arrive long after you and your children have passed on.

• That nature produces best out of the great variety of plants and
animals, not out of one-sided systems, or monoculture.

• That nature needs animals and animal manures to keep the fertility of the soil.

• Hundreds of technical skills necessary for skillful use of tools and machines.

All these basic teachings our educational systems and our mostly urban and suburban environment do not provide for modern children. Most children grow up in a way that does not acknowledge or harmonize with the rhythms of nature, or even the rhythm of their own bodies. They grow up in a way that ultimately convinces them that money brings good, and that pleasure in and of itself will bring happiness rather than the work we do for others. Out of this education, they follow their uncontrolled desires, rather than responding intelligently to the necessities of a natural environment. They think that everything is available if only one has the money to buy it, and that the world has never-ending resources. They often see in their surroundings that people live well without contributing anything to society. They cannot experience that the critical difference between industry and agriculture lies in the fact that nature can produce out of its harmony and variety without major input, while industry cannot.

In these conditions young people miss the basic social experience that comes from recognition that in cultivating the earth and caring for animals and plants, one must rely on the work of others who cultivated before you, and that you do not necessarily reap what you have planted, but that others may benefit from your work.

Can the farms of tomorrow again, and in a new way, provide these teachings for young people? If we see the disasterous state of our educational systems now, with their overemphasis on intellectual faculties and their incapability of helping young people to create an inner morality, of directing their will forces creatively out of a strongly developed personality, we can recognize that we have to do everything to bring the farms of the future into a condition where they can directly contribute to the inner and outer development of young people.

We cannot move back to a rural society. We have to create a new relationship between the citizens and "their" farmland that will make the benefits of farm experience available for anyone who seeks education, recreation, or therapy. Every community needs to incorporate farms not only to have fresh local food, but also to have available these educational facilities. The extermination of farms on the East and West Coasts of the United States is leaving a vast, thinly populated, highly mechanized and chemicalized agriculture in between. This has many detrimental consequences, environmental, educational, and social.

Every school, public or private, needs a farm or a group of farms to give students the opportunity for practical training in nature. Therapeutic communities with handicapped, retarded, and chronically diseased people and with juvenile delinquents should also be placed inside or adjacent to healthy farm organisms, and make use of them. Likewise, elders should have the opportunity to work in gardens or farms. In fact, to compensate for the soul-draining work of factories and offices, such opportunities should be available to all.

This will become possible only if farms are diversified, integrated, and oriented toward quality instead of profit. Many existing biodynamic farms have long experience with such educational and restorative efforts, serving as the haven for whole classes of retarded and elderly people, as well as young families with small children who just come to look and play. Naturally, many implications and difficulties can arise out of such situations, but they are resolvable. For the farms of tomorrow to undertake this broad educational and therapeutic task, they will require new generations of farmers who have not only mastered the farm techniques but who are also strong in the humanities and in social skills. Healthy food, environmental care, and educational possibilities — for all these essential sources of life and for inner development, we need the new farms.

ESSAY 2
What Is Needed To Create The Farms Of Tomorrow?
Concept — Land — People

The farms of tomorrow must arise from a new concept, a new leading idea that serves the basic aims of agriculture. Those aims are, first, to grow life-filled, health-giving food in ample quantity and diversity to feed the local community and to serve regional and urban needs that are not met locally; second, to do this in a way that not only conserves but improves the natural environment; and third, to give all who want it the educational experience of working with nature. Without a leading concept that concerns itself with the wisdom that lies in nature and with the relationship of the human being to nature, we will be unable to create new farms that will serve these three purposes.

In the past, the motivation for agriculture was primarily taken from the need to support oneself and one's family with food, firewood, and clothing. The methods of farming were shaped by experience and the traditions that resulted from them. Far into the 18th Century, farming was not so much an economic venture as a means of self-support, and also the general lifestyle. In that sense, it was pre-economic. Before industrialization and the growth of cities, most people were engaged in farming. There was no real market for agricultural goods. For many centuries the only money that was needed was money to pay taxes to support the nobility, their soldiers, and the clergy who did not support themselves through farming, and also to buy necessities such as tools for farming and salt. Salt was essential because in cold climates one could not survive the winter without salted meat, fish and vegetables. To get the little necessary cash for these things, many people went into a craft or a service business without giving up farming. They became blacksmiths, carpenters, or innkeepers in their home villages, and by this created tradeable goods or services. So the general pattern was for rural people to keep farms to feed themselves, and produce goods or render services to trade. Only toward the end of the 18th Century did farming, very slowly, become a business itself. It was in this epoch that agronomists like the German Albrecht Thaer proclaimed "agriculture is a trade, the purpose of which is to make profits or money. Farming is a way to earn money like any other business."

The motive to earn money through farming, to make a profit — profit being the difference between money input and money earned — took its place beside the traditional values of farming, and steadily became more and more domineering. The rapid development of natural science in the 18th and 19th Century, and the concurrent development of agricultural science, provided the tools for a vast and necessary expansion of agricultural production. Modern agriculture was formed through the combination of the new economic approach and agricultural science with the rapid growth of population and the expanded economic resources available through industrialization.

Agricultural science took more and more to the new economic trend. It aimed less at exploring the ideal conditions under which a whole farm with its plants and animals thrives as a natural organism. Instead, science turned the art of agriculture into agronomy, techniques of exploiting soils, plants and animals for monetary profit. The guiding question of agricultural science has been, under what conditions is plant or animal production the most profitable — with profit measured solely in money. The nature of the farm organism and the question of its relationship to the environment was rarely considered.

Generally questions of quality became, and still are, secondary to questions of profit. If we look at the farm scene of America today, at the farm crises of this century, at the devastating impact of this approach to farming and to our natural environment with its vanishing soil, its sick and vanishing forests, its polluted ground water and its often miserable rural population, we perceive what a high price not only the rural population, but the whole of society has to pay. It has become obvious that the profit motivation does not lead to healthy life-giving food, nor to conservation or improvement of the environment. The history of agriculture in the last 200 years proves this clearly. As stated in the first essay, we need farms for three reasons: for healthy food, for a healthy environment, and for cultural and educational reasons. In dealing with these needs we have to be aware that they are basic to everyone, and in creating the farms of the future we have to make sure that the needs of all

18

are met. Consequently, three different motivations have to come together to shape the farms of tomorrow.

• The first is the basic spiritual motivation: that every year life on earth is created anew, so that human beings can be born safely and have healthy bodies that will allow them to live out their individual and collective spiritual destinies.

• The second is a social motivation: to shape our land use with the goal that everyone have access to healthy food, wood, and fiber in the right amount and independent of his or her life situation.

• The third is the economic motivation that makes all other goals possible, and is the basis of the new farm concept. We must develop the farms of tomorrow in such a way that they regenerate themselves more economically and become more and more diversified, serving as the primary source of food for the local community. This diversity and regeneration should arise with the help of the forces of nature inside the farm organism so that it becomes less and less necessary to introduce into the organism substances and energy from outside such as feed, manures, and fuels, and so that human labor is used as economically as possible. Stated another way, the economic ideal is a farm that achieves and maintains high fertility within itself, generating a surplus of food for the community, and its own seeds for the coming year while the input of outside substances, energies, and labor goes toward zero.

This truly economic motivation should not be confused in any way with the profit motivation. They are totally different categories. Many things we do today in farming are profitable but uneconomic. For example, nowadays strawberries are frequently grown in California for the Northeast market, even when those berries could be grown in the Northeast itself. The grower, the trucker, and the retailer will all eventually make a profit, but there is a hidden loss. The amount of energy expended to grow and transport the strawberries to market far exceeds the amount of energy they will yield when they are consumed. There is a far higher input than output. Ultimately, society must cover this loss in some way. Thus, the profit of a few becomes the loss of many. Production is truly economic when it is done with the lowest possible input of substances, energy, and labor, and when the output exceeds the input. As we enter an era of dwindling resources, many people are recognizing the need for the farms of tomorrow to be truly economic: to renew themselves while creating a consumable surplus without the input of substances from off the farm, and with a reasonable expenditure of labor and energy.

A Leading Concept

There is a well-developed agricultural concept that can guide us as we work to create the farms of tomorrow. This leading concept was given to European farmers in 1924 in eight lectures by Dr. Rudolf Steiner, the Austrian philosopher and natural scientist. This concept has come to be known as biodynamic farming. It has been put into practice over the last 65 years on hundreds of farms on all the continents. In America it took hold mainly through the work of Dr. Ehrenfried Pfeiffer. Biodynamic farming serves predominantly the motivations mentioned above. Over the years it has reliably demonstrated its capacity to lead farms to high quality on all levels, and to the most developed economy.

The idea of the farm as organism is widely unknown today, yet it is of high importance for those farms that work with the advice of Dr. Rudolf Steiner. In the biodynamic approach, the farm is seen as an organism, and that underlying concept is part of all considerations and actions. By definition, an organism is a living entity consisting of parts, or organs. The function of each part is essential to the existence of the whole, and also to each of the parts. An organism has its own inner life and circulation that is different from its surroundings. An organism develops over time, having a beginning and an end and its performance depends on the harmony of its parts, or organs.

More and more scientists are beginning to perceive the whole earth as an organism, thanks in part to the work of Dr. James Lovelock and the inspiration of his Gaia Hypothesis, Because of the essential exchange processes of oxygen and carbon dioxide, the earth's destiny depends on a harmonious ratio between mankind with the animal world and the plant world. As everybody can see now, we need the right ratio between forest and open land; we must have, for example, the oxygen-producing and water-conserving belt of rain forests.

So in this context arises the question, how can a farm be or become an organism in a true sense? Every farm and every organism lives in space and has boundaries. There are legal boundaries, such as the title, which defines the land which can be part of the farm; but legal boundaries are abstract and create no living organism. In nature, space is created and formed by the animals. We are living in a network of animal territories. Wild animals mark and defend their territories; naturally these territories are not related to our legal boundaries. Deer, for example, move over a large territory not observing any legal boundaries.

The organism farm is created by the domestic animals. Inside the farm's legal boundaries the farmer establishes a herd or various herds. Ideally, they feed from the vegetation inside the farm, summer and winter, without anything brought in from other farms. The animals respond with a manure that is formed exclusively by the flora of the farm organism. This manure is collected, and when properly treated comes back to the plants of this place, stimulating them. In this process of correspondence between farm animal and farm vegetation, the farm develops and becomes more and more individualized. Over time the animals adapt and become rooted to their place. In the intestines of the ruminants a special ferment pattern develops that is adapted to the flora of the place. This is reflected in better health of the animals and in a better performance. For example, early tests on farms organisms have shown that after turning to biodynamics cows use less food protein to build up a pound of milk protein. The animals "make more of their food."

The inner health of the animals radiates back through the manure into the plant world. An organism is born and develops in time. As with any organism, the farm organism needs a strong inner circulation of substances. Depending on the quality of the soil, the farmer has to determine what percentage of the acreage can be used for market crops and what part has to serve the inner circulation through green manure and fodder crops. The condition and the amount of organic matter or carbon substance are decisive factors for the health and the productivity of a farm.

Free Land

For the farms of tomorrow, land cannot be used as a commodity or a tradeable good, like a car or a pair of shoes that are produced, sold, used, resold, and finally used up. After all, land lacks every attribute of a tradeable good: it has not been produced, it cannot be produced; it is limited in size to the surface of the earth; and access to it is essential for every human being. The basic economic reality is the relation between the population of a given region and the amount of usable land in that region. No one should possibly have more of the fruits of this earth than what grows on the amount of land that is arrived at when you divide the amount of usable land of a region by the number of people living there — and no one should have less.

The widely held belief that there is such a thing as private ownership of pieces of our living planet is a fiction, a social lie. As any real estate attorney would agree, when one holds title to a piece of land, one actually holds a bundle of rights: the right to use a piece of land exclusively or in cooperation with others unlimited in time, and the right to hand these rights of land use on to successors. If you buy and sell land, you are actually buying and selling rights of use, not a commodity, which land realistically cannot be. Now, the buying and selling of rights and titles is a dubious practice with an odious history.

In former times it was the right of certain families to hold designated public offices. That right was considered inheritable. The Kings of France, with their assumed right to raise taxes, assigned certain families to collect vast sums of money to finance the court's excessive lifestyle. So the right for taxation went into private hands and was executed there for the profit of these individuals. Millions of people died of starvation in rural France by overtaxation. Another odious practice of trading rights is to achieve the right to hold a certain office — ambassador, consul, and so forth — by contributing heavily to the campaign funds of political parties.

The right to draw revenues from spiritual production — the copyright — has been limited in most countries to a period of 30 or 50 years. So there is a growing feeling that rights should not be traded limitlessly. In the future this will apply more and more to land titles as well. To give a common example: a family had farmland in use over a long period. They stopped using the land in, say, 1940 and moved out of the area to live their old years somewhere else. They keep the title to this land. They and their children do not care for this land for 50 years. By holding the title they exclude anyone else from the use of the land. The land bushes in, becomes an impenetrable thicket of no pleasant appearance, and eventually a quasi-forest of very low productivity. Suddenly, and without any help from these titleholders, an economic boom reaches the area. Land is needed for development. The titleholders sell their title, making an enormous profit that is raised on the labor of those that are in need of the land now.

Hand in hand with the fiction of private property versus parts of living nature, we developed the use of land as collateral, the mortgaging of land. But the history of farm mortgage is the history of farm crises, the depletion of land, of deforestation, or erosion. In its 1938 Yearbook of Agriculture, the U.S. Department of Agriculture wrote: "Mortgage may be as injurious to a farm as erosion or a poor cropping system," and "it has often been pointed out that certain tenant agreements result in overcropping and depletion of the soil but it is not always recognized that this same condition may be caused by the terms of a mortgage on a farm that is operated by its owner. The necessity of meeting payments on a mortgage . . . has caused many farmers to specialize in crops that ultimately reduce soil fertility, to neglect the restoration of humus, and to fail to plant crops that prevent erosion. At the same time the farm house and the other farm buildings as well as the soil deteriorate."

Property taxes place the same kind of negative pressure on farms. Whether the farm has a good year or bad, whether there is drought, or hail, or general crop failure, both mortgage and tax payments come due. They must be paid or the property is forfeited. This again forces farmers into growing market crops that promise a short-term profit. The USDA report of 1938 points that out very clearly: ". . . too great reliance on property taxation in rural communities tends to promote short-sighted land use, which if persisted in, brings about serious deterioration of the land resource."

As the socialist and communist nations of the world have demonstrated unequivocably, state monopoly of land is equally disasterous. Government control destroys individual initiative and with it the land itself. For the farms of tomorrow to succeed, any form of government ownership or even administration of land has to be strictly avoided. Government is unable and unsuited to deal with this task.

The farms of tomorrow must be based on a new approach to land. The land can no longer be used as a collateral for debt, it should no longer be mortgaged. It must be free to serve its original purpose: the basis of the physical existence of humanity. How will this be possible under a Constitution that sees land as a commodity and protects the property rights of the title holders? The answer is straightforward. The land has to be liberated out of the insight and actions of citizens who recognize the essential need for "free" land. Specifically, local land suitable for agriculture must be gradually protected by land trusts (see Appendix C). To do this, every piece of farmland has to be purchased for the last time, and then, out of the free initiative of local people, be placed into forms of trust that will protect it from ever again being mortgaged or sold for the sake of private profit. Non-profit land trusts must then make the land available to qualified people who want to take it into ecologically sound uses. Such arrangements will give the right of land use to individuals or groups, either for the time they are willing or capable of using it, or in a lifelong contract that could even include the right to find one's successors.

The U.S. has a growing land trust movement. These trusts take land into their legal embrace so that it can serve the two basic needs of humanity — ecologically sound farming systems and affordable, non-speculative housing — in a way that excludes profiteering. It is neither necessary, nor likely, nor desirable for legislatures to change Constitutional or other legal rights relating to property. The soundest way into change for the better, is for property owners, out of an understanding of the necessity for free agricultural land, to gradually and freely donate or sell their land into such trusts. Landowners themselves could form such trusts, or groups of citizens could cooperate locally to buy the available land for ecologically sound farming and for affordable housing projects. This is clearly something that cannot be legislated or otherwise imposed in any way upon humanity. Every step of progress will have to arise out of the insight and the free initiative of the people. Our expectation towards legislators and government administrators can only be that they will not hinder such development, and perhaps further it in some ways.

24

The transfer of more and more land from private ownership into nonprofit land trusts will raise many serious financial questions. For example, if the land cannot be used as collateral by the farmer, if it cannot be mortgaged, new ways of farm financing have to be opened up. These new financing mechanisms will be based more on personal credit. This personal credit will be obtained more easily as more people are responsibly connected to a farm operation. In this case financing can arise from guarantor communities of people who are directly connected to a local farm project, as with many of the community supported farms described in this book.

Another great question arises out of the fact that, by and large, in the U.S. today a farmer's financial safeguards for old age and illness are based on the market value of his farm. Against this market value, farmers in need can borrow money and in the end, when they sell the farm, finance their old age. If a new way of financing these needs cannot be found, a new approach to land property and land use would be difficult. In a community supported farm system the cost of illness and old age of the farmers can be met through all the household incomes of those households that are connected to the farm. They become a part of the operational cost of the farm. This can be done with or without the conventional insurance techniques.

The land question is closely connected with the social welfare question in general. For most people, their only source of financial support after retirement is directly linked to the real estate they own, usually their home or their farm. The questions of financial safeguards against illness and old age go far beyond the farmers. But to heal the many social problems we have now, we must begin with the farmers. They, after all, care for the fundamental source of renewable wealth: the land.

Furthermore we have to take into account how much our whole system of money and financing today depends on our mortgage system. The "miracle" of economic growth with its credit financing not only of investments but also of consumer goods, in the way of "consume today, pay tomorrow" is, besides giving credit on income, mainly based on mortgaging land and its buildings. Consequently, the volume of credit can be expanded insofar as the market prices for land and houses rise. But whenever the real estate market stagnates or declines, we inevitably witness the high indebtedness of the farms leading directly to a massive crisis with thousands of farm foreclosures (1928-35, and 1986-?). This is clearly an unstable and inherently dangerous system, which causes tremendous hardships for millions of people. The financial system, especially in regard to mankind's essential agricultural activity, must be restructured. Our present money system has, to a large extent, led to the ruination of our farms, our food quality, and our environment.

Ultimately, we must develop a nationwide system of "free land" for agriculture, as described above. This system of free land that cannot serve as collateral will, obviously, imply a different money system altogether. That is one of the critical challenges of the future.

People

In olden cultures such as the classical Greek and Roman, work on the land was generally done by slaves. In Greece this slave labor was provided by the original inhabitants of the land who were overcome by the Dorians invading the peninsula from the North. In Rome the labor was done by various subjugated peoples. The dependency of the peasant population and their slave-like status in society continued far beyond the Middle Ages. In Russia, certainly, it continued into the 19th Century, and the same can be said of sharecroppers in America. More and more through various reforms, dependent peasants became independent farmers. But these legally independent farmers soon came into the dependency of financial institutions. The very act of acquiring a farm through mortgaging led directly into this new dependency. The farm laborers became hired personnel. They had to sell their labor for money to make a livelihood, and they still do today.

The farms of the 18th and 19th Century in North America were carried usually by large families and the help of hired — often migrant — labor. With the concept of the family farm we connect still certain emotional, moral, and social qualities. So saving the family farm is often a declared aim of politicians and agricultural writers. They cling to this model, even as it declines in the face of more and more agribusiness — capital corporations that manage plant and animal production through hired management and labor. The defenders of the family farm seem often to forget that what they idealize was in the past mostly based on large families, often three

26

generations, working together on one farm, substituted even by hired labor. Unmarried sisters and brothers often stuck to the farm and were welcome and cheap labor. This clan-like family structure is no longer prevalent. The family farm of today means usually one couple managing a far too big land mass and far too big animal operation by themselves and at the same time raising a family. Usually the women in these operations are totally overburdened. Hired labor, no longer available nor affordable, is replaced by machinery. With machinery comes debt and mortgages; the freedom to do what is right on the farm vanishes. So the family farm can no longer be the ideal.

Experts estimate that 25% of the usable land in the U.S. will be managed by agribusiness by the year 2000. The alternative to this trend must be a new association and cooperation of families and individuals, excluding the use of hired labor. The typical attitude in farming is a complete devotion to the cause, to the work. If you are not totally dedicated to the animals and plants, to the whole farm organism — without any reservation — the animals and plants will not thrive properly. The hired hand with his wage contract cannot develop the necessary devotion.

In some community supported farm systems, member households carry a certain part of the budget out of various sources of household income. The life needs of those who principally work on the farm, and who have no outside income, are supported by the household incomes of the other farm members. That means that the farmer's family is not driven to make a profit with which it can support itself, nor is the farmer hired for a wage to work for the others. He has been brought into a position where he can donate his labor out of his spiritual intentions for the well being of the farm organism and the fellow members of the farm community. Beyond that every other member family of this community supported farm can come occasionally, or regularly, to donate some work for the benefit of the farm and the community. Beyond this, donating work to the farm benefits the community member by giving him or her the experience of working in nature and the attendant opportunity to deepen an understanding of nature and of him or herself. In such a situation people can experience the possibility of a new relationship to labor. And in considering this possibility, we can gain a sense of how humanity develops in its relationship with labor: from slave labor, to hired labor, to donated labor.

The cooperation of free individuals on free land to ensure the basic needs of all has another ideal background. As stated earlier, land use is a basic necessity for every individual. Without the right to use land, no individual has security. Since the amount of usable land in an area is constant, and the size of the population at a given moment is also known, we can exactly determine at any given time how much usable land is available per person in a given area (in Central Europe about half an acre per person, in America about three acres per person). This ideal/real relationship between land and people is a basic feature of our economy.

The right of the use of a certain amount of land for every individual can be supplemented by the responsibility of every adult for a given area. This right of land use and this responsibility can be shared. As shown in some of the examples in this book, people can associate to share their rights and their responsibilities, and so create a true farm organism wherein they will share the care and the cost of the operation. This can be one theoretical basis for community supported farms now, or on the farms of tomorrow, where any given community supports itself through farming.

ESSAY 3
Ten Steps Towards The Farm of Tomorrow

Many other authors have written extensively on the concept of biodynamic farming (see the Recommended Reading list). Here, in brief, we outline ten basic steps that underlie this farm of tomorrow.

1. The first step for farmers and gardeners should be to remain in the realm of the living with all measures and applications. This excludes most all mineral and synthetic substances for use on plants, the soil, or animals, including mineral fertilizers, synthetic pesticides, herbicides and fungicides, and mineral supplements in animal feeds. Life processes can only be generated out of substances already filled with life. Soil fertilization should always mean enlivening the soil with living substances: manure, compost, and all sorts of green manures. Minerally fertilized soils lose their life capacities (which can be easily measured in soil bacteria per cubic inch), and their capacity to store water and air. As a consequence of this, more and more energy must be used for mechanical soil preparation and for water soluble mineral nutrients.

An exception to this rule comes in the use of lime, such as ground limestone or Dolomite, on acid soils. But even here experience has shown the lime substances that are closer to the life process, such as ground seashells and calcified algae, are more effective; smaller amounts can be used than in the case of ground limestone. But if you do use ground limestone, it can be better and more effective to lead this mineral substance through the composting process by adding a certain amount of it to the organic composting material.

In animal husbandry, mineral supplements to the feed can be almost totally avoided if we provide the animals with a wide spectrum of plants. For the ruminants these should include the availability of perennial plants like bushes and trees with their leaves. Such perennials, together with herbal plants, commonly have a higher concentration of mineral substances in the finest preparation. Generally herbal plants of the families labiatae, compoitae and umbelliferae, should play a larger role as feed supplements. One exception to this rule in animal feeds is salt which

29

might be necessary as a supplement out of the mineral world. Pigs can be kept free of mineral feed supplement if they are kept in contact with living soil with its rich antibiotic fungus flora.

Experience has shown that there is a great advantage of always staying with measures in the realm of the living. In using in field and farm only substances that are derived out of life processes, one can hardly ever overdose or create any harmful effects. To overdo the application of compost in the field can be wasteful, never harmful. The same is true for plant-derived sprays which may be used to regulate insect attacks or fungus diseases. By contrast, just think what immense poisonous effects appear from the wrong application of synthetic herbicides, fungicides and insecticides. Mark the rule: in all farm or garden measures and applications, stay in the realm of the living.

2. The second step is to arrive at the manure that is necessary for healthy plant growth by keeping on the farm a sufficient number of animals in the right harmonious combination of species. One can have a balanced farm organism keeping mainly cattle, because of the harmonious characteristics of its excrement that relate in a special way to the fertility of the soil. However, it is better to have a mixed population of different animals that include birds (chickens) and pigs, eventually horses and sheep or goats. The character of each animal manure is different. While cow manure is very balanced with an emphasis on calcium, pig manure has a potash character, and bird manure has the character of phosphorous. These specialties of the different manures can be used in special plant cultivation. Generally, it can be recommended that the different manures be mixed and processed together.

The quality of animal manures depends largely on their diet. Only a mixed diet that is specific for any species gives us a manure quality that serves the soil in a way that will produce the best plants. The time of passage of feed through a cow can vary greatly. However, protein-rich feeds and silage reduce the passage time, and lead to a manure that behaves differently in fermentation. These often smelly and more liquid excrements attract flies and decompose differently. If ever you have grazed cattle and sheep in pasture that includes bushland and forest, you can observe the totally different quality of excrement.

A broad and varied diet that is supportive of the various animal species is the basis of a balanced manure and, consequently, healthy plant growth. The right number of animals and the right mixture of animals has to be found by observation of the individual situation of the farm. Roughly, one can expect that one cow (1,100 lbs.) needs two acres for year-round forage, and can keep four acres fertile by its manure. Of the mineral substances our domestic animals take in with their feed, 80% is recycled through the manure into the farm.

3. The third step is to feed the herds — all the animals on the farm — from feeds that are grown on the farm organism itself. This step allows the farm to build its strength and individual character as an organism. The exchange of the feeds from the farm organism to the herds, and from the herd's manure back to its soil, creates a process of mutual adaptation. This adaptation finds its expression in the microbiotic life in the digestive tract of the animals that gets shaped by the specific flora of the farm (a fermentation pattern), and leads to higher performance and a higher level of health in the animals. One expression of this performance is the capacity of the animals to bring forth new life. The average number of calvings in most industrialized countries is between two and three calves per cow-life; the average in a well-developed farm organism with a well-adapted herd should be six to seven (of an animal that can, ultimately, come to 15 calvings per life).

The aim and the result of this local interchange between animal and forage is permanent fertility. In keeping the necessary number of animals on the farm, we necessarily introduce more fodder crops into our rotation. These fodder crops, such as alfalfa, clover and grasses, provide a great root mass for the upkeep of the organic substance in the soil, and help create a soil based on life, not fertilizers. A great variety of fodder plants should be used to serve the needs of the animals for a balanced diet and to create balanced crop rotation that keeps up the healthy fertility of the soil. Practically no feeds should come from outside the farm organism.

4. The fourth step is to aim for a great diversity of plants on the farm in combination with, and as part of, the crop rotation. Fertility

and productivity in nature arise out of diversity, not out of specialization or monoculture. The author counted up to 70 plant species in use on one of the biodynamic farms he served, including six grains, eight leguminous fodder plants, 12 grasses, numerous brassicas, vegetables, and herbs. By contrast, a modern dairy farm in New England cultivates, beyond its hayfields, usually only one species: corn.

A well-managed pasture should have, besides grasses and clovers, deep-rooting plants like alfalfa and herbs. They mobilize in the soil mineral substances from lower levels and make them available to other crops. The management of such pasture should be so that the root growth of the plants is not inhibited by overgrazing late in the year. Plant rotation in the fields can be looked at as a diverse plant community spread out over time. Here again, deep-rooting plants that leave behind masses of organic material should alternate with shallow-rooting plants.

For the feeding of mankind, the total production of our farms counts more than a high level of yields in special crops. Experience with many organic and biodynamic farms has proved that their total output of food, measured in calories (Joule) is generally higher than the output of conventional farms, where the yields in their single market crops can be substantially higher. The reason for this is in the diversity of crops on biodynamic farms and the productivity that results out of this diversity. Plant species support and complement each other more than they hinder each other.

5. The fifth step is to recognize that the circulation of carbon, or organic substance, throughout the soil, the plants, and the air is the basis of permanent fertility. This circulation expresses itself in the creation and breakdown of humus substance in the soil. Consequently, great care must be given to carbon circulation, the proper breakdown and buildup of humus. Experts estimate that through the last 150 years, 50% of the humus substance of the earth has been lost — either used up, or washed away through erosion. The life processes of the soil microorganisms, earthworms, springtails and so forth depend on the functioning of this living humus layer of the earth. And humus is built up by animal manures and plant residues (root masses and green manures). To keep soils from being

exploited, turnover of an estimated 30,000 lbs. of carbon matter per acre per year is necessary and sufficient. To achieve this we have to find out for each farm what amount of the crops can be sold and what amount has to be green manures that will go into the humus circulation. The acreage of the farm that can carry a market crop will necessarily be smaller on a poor soil than on a fertile soil, versus the acreage of crops grown for green manure.

An essential tool in keeping up the humus level is the right or sweet fermentation of animal manure, and the composting of plant matter. To guide these fermentation processes in the right way one can use herbal composts that are implanted in manure and compost piles. These special herbal composts can be prepared by the farmer, or purchased from Biodynamic Preparations, Inc. (see Appendix B, Resources). Another great factor in the care for the humus layer is cover cropping and adequate tillage. If we have nutrient deficiencies on an organic or biodynamic farm, we do not work on the nutrients directly by applying them, but instead we work on the humus level and the humus quality.

6. The sixth step is to strengthen silica circulation in the soil by encouraging microbiotic processes. As we deal with the carbon cycle and with the life forces and the humidity in the soil, we also need to pay attention to the circulation of silica, which is linked with the light and warmth forces in the processes of nature. The Russian scientist V.T. Vernadski declared "No doubt is possible about the fact that no living substance, no organism, can exist without silicium."

The ashes of many trees and of the high grasses contain large amounts of silica. In most other plant species, silica plays an important role in the roots. In spite of this, silica can be called the forgotten element in agriculture. Among the 15 elements that are most frequently named as essential for healthy plant growth (Carbon, Oxygen, Hydrogen, Nitrogen, Sulphur, Phosphorous, Potassium, Calcium, Magnesium, Iron, Boron, Molybdenum, Manganese, Copper, and Zinc), Silicum is not mentioned.

One reason for this is that silica is not solvable in water. It enters the life process of plants only through the living activity of soil microorganisms. If enough silica is taken in by the plant, more happens in the plant through the influence of the sun because silica transports light and warmth. Both directly and indirectly, the sun forces and other cosmic influences in their different qualities are transferred to the plant through the silica. The parts of the plant that contain much silica grow twice as fast as the plant organs that contain more calcium (Voronkov, *Silicium and Leben,* Berlin, 1975). One of the problems of one-sided fertilization of plants is that it changes their mineral composition drastically by reducing their silica content. As chemical analysis of plant ash has shown, between 1840 and 1940, the silica content in plants has been reduced drastically, and we can assume that in most commercial agriculture the downward trend has continued into the 1990s:

	Wheat Straw	Rye Straw	Barley Straw	Oat Straw
1840	81%	82%	71%	79%
1940	63%	48.2%	52%	46.5%

(Duftos, 1840 and Menzel Lengerke, 1940)

In consequence of this reduction in silica, the straw loses its resistance and strength, becoming prone to fungus infection. If the silica content has been reduced in the straw, then one can also assume that a similar reduction has taken place in the grains eaten by livestock and human beings.

High intake of mineral salts, caused by the use of petrochemical-based fertilizers, in all probability in turn causes the reduction in silica content. The reduction in silica content parallels a heightening of potash content. Modern agriculture tends to push the growth of plants with the use of nitrogen-rich fertilizers, but this changes the inherent balance of substances within plants and frequently makes them prone to pests and diseases of all sorts. The crystalline substance silica is essential for healthy plant structure, and also for human and animal health. But silica can only be absorbed gradually into the plant through the aid of the microbial life in the soil. For that reason, rather than feeding nitrogen-rich fertilizers to the plants themselves, the farms of tomorrow must gradually encourage the life in the soil with a broad and balanced supply of organic matter. As the microbial life of the soil thrives, so the plants that arise from the soil will be able to take in a balanced quantity of silica and so thrive.

The use of finely diluted and specially prepared crystal silica promotes balanced growth processes in green plants.

7. Step 7 is to create harmonious balanced conditions in soil, plants, animals, and landscape as the necessary basis of productivity. The health of plant and animal organisms is a function of this harmony.

Out of the six previously mentioned steps we can begin to see that productivity in nature arises out of a harmonious, balanced situation on all levels. This situation we have to achieve in the landscape, in the number and mix of domestic animals on the farm, in the building up and the breaking down of humus substance, and in the metabolic processes in the animals and the plants.

There is a polarity of warmth and light processes to water and earth processes, and likewise a polarity of warmth and light substances to water and earth substances. In testing a wide variety of plant material, the great American soil specialist Firman Bear found that every species has a specific relation between these substances, and that this establishes the pattern of the plant.* The plant is not healthy and nutritious because it has an abundance or deficiency of certain substances, but rather because it has all of its substances in harmonious proportions and order. There has to be kept a certain balance between the substances which build base and those which build acid. The balance is typical for each plant, and it can easily be destroyed by unbalanced mineral fertilizing, making the plants prone to insect pests and fungus diseases.

From the landscape we have to work toward a balance between forest, field, pasture and wetland down to the inner structure of the plant. We must work and produce out of harmonious balanced situations.

8. The destroyed natural environment must be restored. In particular, for this restoration to occur we must pay attention to two elements: hedgerows and wetlands.

---

* (Firman Bear: *Soil and Fertilizers, Cation and Anion Relationship and Their Bearing on Crop Quality*, Rutgers University, Department of Soils.)

The hedgerow is the natural habitat of many valuable birds and insects, a natural barrier against wind erosion, and also an invaluable aid to maintaining humidity in the soil. Birds and other creatures which find a home in hedgerows feed upon harmful insects and so help to maintain the natural balance. Valuable insects also live in hedgerows and on the fallow land beside the fields. Ant colonies, for example, often establish themselves in hedgerows. The ants may range up to 200 feet from their nests, and are a reliable control for the larvae of the potato beetle. Hedgerows also provide a source of wild fruit for man and animals. By nibbling on the hedgerow, cattle diversify their diet and so maintain optimum health. Hedgerows should not be established as monotonous lines of trees, but rather should consist of a great variety of bushes and trees so they will attract a great variety of birds and insects. Many blossom- and fruit-bearing varieties should be incorporated into the hedgerows, such as quince, filbert, elderberry, and fruit-bearing roses. More and more people will in the future want to supplement their diet with wild fruit.

Another healthy factor in the landscape of a farm organism can be ponds and wetlands. They enhance the production of dew, which supplies water to the plants during dry periods. They are also the habitat of amphibians, and of many insects and birds which are beneficial for the environment. The moist environment of the wetlands attracts microorganisms, insects, and a mushroom flora thereby distracting them from the croplands. Consequently, with ponds and wetlands again we can expect and observe a regulating effect in the whole ecology of the farm.

9. Step 9 is to implement biological weed and insect pest control. As described in Step 8, a sound diversified ecological system that contains forest, field, permanent pasture, hedgerows and wetlands, with numerous domestic and wild animals, is the basis of biological weed and pest control. A second basis is to develop healthy humus content in the topsoil. The antibiotic fungus flora of a developed humus has a cleansing, detoxifying influence on unprocessed, or raw organic substances. To keep certain insects under control it is nevertheless important to bring very little unprocessed organic substance, such as raw manure, into the soil. The most effective

procedure against weeds and pests is a diversified plant rotation in the fields. Finally, over the last 65 years many steps have been taken to isolate out of plant and animal organs substances that help to reduce the pressure of weeds and insect pests. Those substances can be used to good effect, without ever having to resort to chemical poisons.

10. Step 10 is to reestablish a rhythmical order in animal husbandry and field care that is connected to the rhythms of the earth and its cosmic environment of the sun, the moon, and the other planets. This is an essential part of farming in the future. Life is rhythm. All life processes develop in rhythms that are related to cosmic rhythms. The moon rhythm is obvious in the gestation periods of man and the domestic animals. The sun force in summer and in winter constitutes different influences on the lower animals and on plant life. Research into these rhythms is just at its beginning, but numerous planetary influences on plant and animal life have been found and documented over the last 50 years. Much of that research can be found in the publications of Maria Thun, especially in her annual calendar *Planting With The Stars*, in the *Kimberton Hills Agricultural Calendar*, and in Lilli Kolisko's *Agriculture of Tomorrow*.

Following the basic principles of these ten points will yield ever greater individualization of every farm. By restricting, and in some cases eliminating altogether, the import of substances foreign to it, the individual farm organism protects its inner harmony and rhythm. The steady inner rotation of substances and seeds, and great diversification, lead to a healthy adaption of plants and animals to the specific micro-environment of the farm organism, and to high productivity. As recently as 100 years ago, the tiny country of Switzerland produced over a hundred varieties of wheat — each variety specifically adapted to the micro-environment of one of its valleys. Today, by contrast, only about a dozen varieties of wheat are grown worldwide, and those varieties need to be replaced every ten years or so because they have become susceptible to pests and diseases. Such a generic, mono-cultural approach not only narrows the diet of humans and their domestic animals, it also makes the worldwide food supply extremely vulnerable.

The economic goal of the biodynamic farm is to have rising quality, and diversification of products which are adapted to the need of the local population. Meanwhile, the input of foreign substances and energies goes toward zero. Such an approach has been proved to be possible in hundreds of biodynamic farms all over the world and is the only real, undiminishing source of support for mankind. While not all of the farms described in this book employ the biodynamic approach, elements of biodynamics are to be found in every one of them.

ESSAY 4
Three Basic Rules

Many things written in this book would appear illusionary and unrealistic if there were not already examples of farms where, in one way or another, some of the ideas mentioned have been realized for some time. Immense social, economic, and legal obstacles lie in the way of structuring our farms in new ways, and restructuring generally and specifically our relation to this earth. However, those who examine the situation will recognize that the crisis of our food includes deteriorating quality as well as insufficient availability for many millions of humans. Realizing this crisis, and the general degeneration of our natural environment, and seeing also the education crisis and the moral insanity of our time, we have not much choice. We must reestablish our relationship to the basic sources of our livelihood; we can do this best through helping to create the farms of tomorrow and bringing ourselves into a new relationship to them.

The farms of tomorrow will not be just family farms. They will be so complex and diversified that they will require the cooperation of many different, unrelated, free people. Such farms will need the cooperation of several households as well as the open support of households whose members are not actively farming but who share the responsibility, the costs, and the produce with the active farmers.

Three basic rules for making such cooperation possible were once given to author Trauger Groh by Wilhelm Ernst Barkhoff, a lawyer who had a deep interest in and personal experience with these questions. Trauger relates them here in his own words and understanding, because in many years of farming he has found them to be true and helpful.

The first rule is, do not work too many hours. Farming is labor, craft, and art. The art arises out of a deeper understanding of nature based on thorough, ongoing observation, reflection, and meditation on all surrounding natural phenomena and processes. If the active farmer is overpowered with labor and is working too long hours, he lacks the leisure for this observation, reflection, and meditation. He loses his art, handing it over to extension services and agricultural schools. They themselves transform the art of farming, which is the true agriculture, into materialistic natural science; and they transform the inherent, developed economics of the farm organism into a merely profit-oriented exploitive agronomy of plant and animal production. Finally the farmer is no more creative himself, he is just applying their recipes. In the end he loses a lot of his craft, too, depending less and less on farm skills and more and more on the supply of sophisticated machinery. He often becomes a badly paid, highly indebted laborer with, hopefully, high technical skills.

The second rule is, buy for the farm as little as possible from the outside world. The less you purchase in the way of sophisticated tools, machinery and farm buildings, the more you are financially independent and free to work with and out of nature. Use the help and the skills of all friends who are related to the farm. The secret of successful Amish economic life lies in relying on human beings to work the farm and not buying too many things. If you work in cooperation with other people — and that is what the farm of the future is about — you will find that purchasing costly tools, machines, and buildings from outside strains the community; it raises social questions. The initiative to purchase these items usually comes from an individual. The consequences of the purchase have to be carried by the group. This raises problems.

The third rule is: Take all the initiative for your actions on the farm out of the realm of the spirit, not out of the realm of money. What does this mean? The more we penetrate the spheres of nature, the more we become aware that what surrounds us in it is of an overwhelming wisdom. What we call scientifically an ecosystem is penetrated by wisdom so that all parts serve the whole in the most economic way. In awe we stand before the higher "intellect" of a bee-hive, or an ant colony. The cooperation of microorganisms, earthworms, springtails, and others in the soil in breaking down plant material and in building humus particles, is something deeply rational and wise which cannot be copied or synthesized by man. This we can describe as the spirit that is spread out in nature.

This spirit organizes nature with the highest economy. For example, the ratio between consumed substance and achieved effect in a bird that migrates from the Arctic to the Antarctic is of a scale that men cannot achieve technically. The comparison between bird flight and an aeroplane demonstrates this clearly. The calcium structure of a hip bone makes it possible to carry a maximum of weight with a minimum of substance, a structure that again surpasses our technical capabilities.

The more we understand and follow this "wisdom" in nature, this outspread "spirit," the more rationally and therefore economically we can organize the farms of tomorrow. The profit motivation, applied to nature, has led to vast depletion of soil and dangerous exploitation of animal and plant material. If we follow the spirit in nature, we put into our service both the rationale and the economy of nature. This, ultimately, is the basis of the life of humanity.

What practical steps can we take now to make the farms of tomorrow possible? We have to bring together, piece by piece, and lot by lot, agricultural land, and then see that it becomes free of speculation and mortgages. We have to establish and support training programs closely connected to farms that are under ecologically sound management, to train and educate the future farmers. And in connection with such farms, we have to establish research programs that support all these aims.

Thousands of men and women in the U.S. own farmland without using it as collateral and without needing to give it over to development. To them we have to appeal, and to those who can give surplus capital to acquire such land. Land given or purchased under this concept must be legally secured so that it can only be used in ecologically sound ways, and so that it cannot be given into development for personal profit. Land trusts have already proved themselves to be one suitable form for this purpose, and a network of land trusts already exists as a supportive model (see Appendices). To prepare the necessary organizations and to start a campaign toward these ends is an urgent need of today.

These aims are only possible if we create, on well-run farms, training programs that enable young people to acquire the necessary understanding for ecological farming, together with the necessary skills. This cannot be done in the form of apprenticeships only. It needs, besides training through work, classroom studies and nature studies with qualified persons. The cost for such training cannot be carried by the farms. Finally, we will have to establish a new way of agricultural research. The task of such research can be defined as

41

studying the social and economic conditions of ecologically sound farming, and also of farming concepts themselves, such as organic and biodynamic agriculture. According to this definition, the new farms described in this book already constitute, in every sense but the formal, an essential research program.

This book was written not to show final results, but rather to point out some approaches leading toward the farms of tomorrow, farms that are needed today.

REFERENCES

Agriculture, Dr. Rudolf Steiner, London, 1974.

Lebensgesetze in Landbau, Dr. Nicolaus Remer, Dornach, 1968.

Organischer Dunger, Dr. Remer, Aunclinghausen, 1980.

The Essentials of Nutrition, Gerhard Schmidt, M.D., Wyoming, RI, 1987.

The Dynamics of Nutrition, Gerhard Schmidt, M.D., Wyoming, RI, 1980.

Silicium und Leben, Voronkow, Zelchan, and Lukewitz, Berlin, 1975.

Biological Transmutations, C.L. Kervran, London, 1972.

Working with the Stars, Maria Thun, 1990.

Soils and Men — Yearbook of Agriculture 1938, United States Department of Agriculture.

Bio-Dynamic Gardening and Farming, Vol. 1-3, E.E. Pfeiffer, Spring Valley, NY 1975.

Soil Fertility, E.E. Pfeiffer, E. Grinstead, England, 1983.

Earth; the Stuff of Life, Firman Bear.

PART II

Community Supported Farms

by Steven S.H. McFadden

EXAMPLE 1
The Temple-Wilton Community Farm

Lean and thoughtful, his arms stretched long by the weight of the milk he carries, Lincoln Geiger makes his way through the barn at Echo Farm. His labor is hard and his body shows it. But he has his satisfactions. He has found a way to work full time as a New Hampshire farmer, and to do it in harmony with the earth. Nowadays, when most food arrives home wrapped in cellophane from parts unknown, Lincoln Geiger is a community treasure.

The good fortune of his circumstances can be traced directly to the tenacity with which he has held to his vision, as well as to the support of about 63 independent families in Temple and Wilton. Together they are creating a community farm that may well be a model for revitalizing the way we grow food. It may also be a model for reconnecting the region's farms to the communities where they are located.

The name of the enterprise, which includes not only Echo Farm but three others as well, is the Temple-Wilton Community Farm. Though they have been in existence only since 1986, they have already brought a tremendous bounty to fruition and they are poised to do more.

As farm member Martin Novom explains it, "our goal is not just to raise food, but also to raise consciousness . . . We don't have all

43

the answers, but we do have some of the questions. How are we going to continue to have sustainable farming? How do we save not just the soil, but also the farmer? Corporate agriculture is not the answer."

The Community Farm

The Temple-Wilton Community Farm is a model that other farms and communities would do well to study, for it may demonstrate a way to revive local agriculture and the plentiful social benefits that come from it.

In January of 1986 Trauger Groh, who had recently moved to the region from a community farm in Germany, met with Lincoln Geiger and several other families who shared the dream of supporting local farms for local people. Dozens of social and economic factors stood in opposition to such an undertaking. Still, they felt they could make their dream real and so they decided to try.

Here's the basic idea they came up with: independent families in the area who wanted to would join together in an association known as the Community Farm. Some of the members had land suitable for farming, most did not. Landholders would make their fields available to those who were able and willing to use them. The rest of the people would continue their lives independently, but through their association in the Community Farm they would receive food from the land. In order to make this possible, they would provide money. (The concepts underlying the development of the Temple-Wilton Community Farm are explored further in Appendix A.)

As Lincoln Geiger explains it, "many people don't want to use the land they have, but they would like to see it farmed. They make their land available so that farmers without land can care for it in their name. Under such an arrangement no one gets rich, but then again, no one goes hungry."

Of the approximately 63 families who belong to the Farm, three individuals in particular work on the land with apprentices. The rest of the families live and work in the surrounding area as teachers, writers, builders, and so forth. They are part of the farm financially and in their hearts, but they all have private lives.

At the start of the growing season the farmers estimate their expenses in detail, including the costs of seed, salaries, tractor repair, diesel fuel, baling twine, and so forth. They present this budget to all the families in the Community Farm — and then collectively they meet the budget. Each family pledges how much it can pay in monthly installments. Families which have much give more than those who have less — not according to a formula, but according to each family's sense of what is affordable and appropriate. They work it out among themselves. Pledges have generally ranged from $35 a month to $80, though the farm membership has debated the advisability of establishing a minimum monthly pledge of $50, which would be $600 a year for a family. These pledges are totalled and applied to the farm's budget.

In 1987, for example, the budget for the Temple-Wilton Community Farm was $53,000. Together the families were able to pledge only $51,500 so the budget had to be cut back a bit. The money gave the farmers working capital and some income to meet living expenses.

Under this model the farmers are not tenants but instead have an ongoing right to use the land. The Community Farm itself owns nothing. All land, equipment, buildings and animals are owned by the members individually. Members who own farm property are compensated only for their costs: taxes, depreciation, maintenance, and so forth. They make no profit, but they keep their land open and productive, the land is improved by the farmers, and the owners serve the community and their families by making it possible to grow fresh, clean food.

To create a land base for their community enterprise, three independently owned farms pool their resources: Echo Farm in Temple, The Temple Road Farm in West Wilton, and Plowshares Farm in Greenfield. Out of about 200 acres total of forest, pasture, and field, they produce nearly all the vegetables for the 63 families who are involved: carrots, beets, onions, parsnips, rutabagas, turnips, lettuce, spinach, chard, Oriental greens, kale, cabbages, broccoli, cauliflower, leeks, celery, tomatoes, peppers, potatoes, squashes, apples, cider, blueberries, eggs, flowers, herbs, milk, and meat. They plan to begin producing their own grain soon.

All the food that goes into the Community Farm storeroom is available to the families. Rich or poor, they take what they need.

If you are hard up for money, you still get food. The community knows you and helps you.

The agreement which binds the families together may be unusual by modern American standards, and it may still have some kinks in it, but so far it has worked. As Trauger Groh sees it, "We need healthy farm organisms not just for the food, but also for our social well being, and for the education of mankind. Farming is so essential that one has to do it at any cost. We can stop making sewing machines or VCRs and life will go on, but we can't stop farming."

Applying Common Sense

Anthony Graham, one of the farmers, says the biodynamic approach they use is much misunderstood. In the main, the techniques are common sense and the application of common sense. To illustrate this point, he explains some of the practices the Temple-Wilton Community Farm uses to grow crops.

They begin the farming process in the depths of winter when the farmers think through their budget for the coming year, and then meet with the families to build a picture of what will grow through the summer. With discussion and active imagination they create the focal points that, in the growing season, they will bring their collective will to bear upon. In support of the farmers, the families pledge to meet all the expenses for the year.

When it comes time for planting, the farmers match the rhythm of their work to the rhythm of the seasons and the planets. For example, biodynamic farmers understand that the phase of the moon is revealing something about its relative ability to influence water on earth — not just the ocean tides, but also the pulse of the fluids in seeds and plants.

As for fertility, compost is the backbone of the operation. Biodynamic farmers don't just spread raw manure on their fields. Instead they compost the manure with straw, other organic matter, and herbal preparations. They use farm equipment to mix the compost so that the materials blend together in a harmonious way. They nurture the microscopic life in the material as it gently cooks over a season in preparation for its return to the soil, part of an unending cycle of birth, growth, and decay.

Biodynamic farmers are fond of spraying their crops, but not with chemicals; they use mineral and herbal preparations. For instance, they spray water mixed rhythmically with specially prepared ground silica, a natural crystalline material which has an affinity for light. Silica allows the plants to work more efficiently, to photosynthesize more effectively.

According to farmer Anthony Graham, the community farm has few problems with insects and disease. That's because they work hard to build good healthy soil, which gives rise to strong plants which can resist disease and pests.

Their next line of protection against insect damage is to spray herb teas. For example, in 1987 when the farm faced an infestation of striped cucumber beetles, which have an inordinate fondness for juicy squash leaves, the farmers sprayed a tea made of pennyroyal. It didn't kill anything, but the beetles found it disagreeable so they vacated the fields for a few days. That disrupted their feeding and breeding cycle. And since the spray wasn't toxic, it harmed none of the insects which prey on the striped cucumber beetle. Consequently, the tea brought the problem to a gentle end without disturbing the vibrant life in the fields. When pests do come, the farmers rely on hand picking, feeding the bugs to the chickens as a part of the farm's internal recycling.

As these examples illustrate, biodynamic farming is labor intensive. It requires much more work than farms which use chemicals to control weeds and pests. Trauger Groh observes that "biodynamic farming needs many hands, many people. You cannot employ laborers because they have gone into industry, and with present economics you couldn't afford to pay them either. Farming does not work well with employees anyhow. It requires a far deeper level of interest and commitment."

Nowadays the Temple-Wilton Community Farm offers places for apprentices, people who are willing to take a step forward into something new, something that could help to preserve the earth and the farms. "Workers don't get paid by the hour," one farmer says, "but then members don't pay by the carrot."

A Principal Challenge

Such an approach to farming calls for dramatic change. But then, if we are to keep our air and water pure, if we are to preserve the remaining farms, and if we are to create a culture which pulls communities together instead of fragmenting them, dramatic change is necessary.

One of the principal challenges of modern farming is, of course, economic. Lincoln Geiger says "if people were to pay the true cost of running a healthy farm, they would have to pay much more. Right now humanity as a whole is paying for it because we are expending resources. We don't pay to maintain the land or the internal economy of the farm. We buy everything from outside and bring those resources to the farm to produce. We need a more labor-intensive agriculture, not a more material-intensive agriculture."

The Temple-Wilton Community Farm found itself facing a serious economic challenge over the winter and spring of 1989. That winter blew in cold and dry on the heels of a scorching summer of heat and drought that reduced crop yields significantly. "It was a poor harvest," Anthony Graham commented, "we had barely enough to get through the winter."

Members of the community also faced the loss of a newly built school, which was destroyed by fire in October, 1988. With the school disaster and the poor harvest in the back of their minds, the membership of the farm came up about $13,000 short when they pledged to meet farmer's 1989 budget of $68,000.

From the beginning, the farmers had emphasized that when you joined the farm you were not just buying vegetables, but rather pledging to support the farm — good years and bad. But when the reality of a small harvest hit home, some families withdrew. And so the farm faced a serious crisis.

In response to the crisis, members of the farm formed a task force and developed several innovative ways not only to cut the deficit, but also to reduce the work load on the farmers. "Now, for the first time," Anthony commented, "it has really begun to feel like a community farm. More and more people are plugging in their various skills to make this work. We feel okay about it. The crisis really galvanized people. We have lots of help now, and we are going to make it."

The Heartbeat of a Nation

For a community farm to work, interest must extend beyond the land into the realm of human relationships. Because of the cooperation required, the members must be willing to know each other intimately and to work out the agreements and disagreements that naturally arise. Such contacts are the grist of community, and they are also what's missing in today's food system: by and large, our relationship with our food and the people who grow it is remote and anonymous.

That there are many people interested in change is evident in the waiting list of families who would like to become part of the Temple-Wilton Community Farm. But the farm is not ready to expand. A farm can serve only so many people. There are limits. For the concept of community farming to grow, other farms and other families must create their own organizations.

As Trauger Groh sees it, that is essential. He says the community farm has no future without a network in New England of 100 or so similar farms that can support each other through trade and association. Such a network, he says, has developed in Europe. "We want to develop a diversified program of foods that meet the needs of local communities and to produce it while the input of energy, materials and labor is reduced."

What the community farmers suggest has far-reaching implications. Agriculture has been the heart of the nation for 200 years, but now the heartbeat is growing weak and irregular. Americans have left the land physically and perhaps spiritually as well. As a result we face grave problems. The community farmers are experimenting to see if they can't begin to show the way back and the way forward at the same time.

EXAMPLE 2
The CSA Garden in Great Barrington

In the Berkshire Mountains of western Massachusetts, farming has become virtually extinct. Aerial photos from earlier in the century show the region was almost completely pastoral, but now, for example, there is only one struggling dairy farm left in South Egremont. Down the road in Great Barrington, the First Agricultural Bank hasn't loaned a cent for farming in over 75 years.

However, without any large pool of capital, a group of about 100 families is demonstrating one way that farming may come back in the region. Together they have formed the CSA Garden. The acronym CSA stands for Community Supported Agriculture, an idea which has enabled the families to directly support some farmers economically, while the farmers produce an abundance of fresh, local food for the families.

Oddly, their story begins not in Great Barrington but in Switzerland. While living there in the early 1980's, a young American named Jan Vander Tuin learned of a new kind of food production cooperative operating in Geneva, Basel, and Vaduz. After studying them, he helped to establish a similar venture, the Co-operative Topinambur in Zurich.

The concept of these new cooperatives is simple: divide the costs of the farm or garden among shareholders before the growing season begins. Instead of an agriculture that is supported by government subsidies, private profits, or martyrs to the cause, they create an organizational form that provides direct support for farmers from the people who eat their food.

Vander Tuin brought the CSA Concept from Switzerland to America. As he shared his passion for the idea around Great Barrington, a core group of people began to form, including John Root, Jr., Robyn Van En, and Charlotte Zanecchia. After the core group got together they contacted gardener Hugh Ratcliffe, and hired him to take responsibility for growing food on a large scale. Hugh had been a research biologist at Cornell University, but he quit when he realized that he disagreed with the predominant scientific view of plant life as basically mechanistic. Instead, he

50

began to study biodynamics. "We can't see the forces at work in the garden," he has noted, "but we can see the effects of working consciously with them."

Biodynamics and CSAs are well suited to each other. While agricultural methods which treat the earth as a living being have the potential to connect people rightly with nature, the community dimension of the CSA garden can likewise connect people rightly with each other. Instead of being confronted with all the usual compromises that a farmer has to make — all the business and marketing issues that have nothing to do with cultivating the earth — under this approach, the support of the community frees the farmers to focus their energy on what they do best: farming.

The First Project

The first project the CSA Garden undertook was in an old apple orchard during the autumn of 1985. They began by selling 30 shares in the orchard, the cost of shares based on the expense of picking, sorting, storing and distributing 360 bushels of apples. They also pressed apples for cider, hard cider, and vinegar. These products were distributed among the 30 shareholders. In the minds of the core group, the success of this harvest established that the CSA concept was really going to work.

While the apple project was underway, the core group leased land for a large garden from Robyn Van En at Indian Line Farm in South Egremont, Massachusetts. The lease was for three years with an option to buy, and it included water, electricity, vehicle access, and the use of an out building.

The group borrowed a tractor, and purchased a disc harrow, a wagon, assorted hand tools, and seeds. They also made arrangements to use a greenhouse for starting seedlings, and a cellar for winter storage space. In the fall of 1985 the land was plowed, composted, manured, harrowed, and seeded with a cover crop.

Over the winter the core group met and estimated that the average person consumes 160-200 pounds of vegetables a year, while the average house size is 2-3 people. Therefore, they figured one share needed to be about 400-600 pounds of vegetables per year, or an average of 10 pounds per week. So production was planned according to these figures, while also considering averages of consumption for 40-50 different kinds of vegetables, herbs, flowers and mushrooms.

Based on their estimates, the core group prepared and issued a proposal for community supported agriculture in the southern Berkshires, and asked for people who supported the idea to send them $10. Only two people did. Still, they felt they could succeed, and so they went ahead with their plans. Eventually they sold about 50 shares to individuals and families that first year, enough to go forward. Each year since then, though, the CSA Garden has grown: 135 shares by 1988, and about 150 shares in 1989 — which is about the garden's upper limit.

The cost of a share has changed from year to year. John Root, Jr. says "when we made our first budget we were worried about too high a price; a share was $557, based on paying every worker $7 per hour. So, in wanting to get the thing to work we priced the wage too low. We have been steadily working it up. A gardener needs to make $40,000 a year to live in the Berkshires. We are only talking about seven months of work, though, so that gets it down in the area of $25,000. In 1988, the gardener made $16,000. In 1989 he will make $20,000."

While wages have gone up, share size and cost have been moderated to meet the needs of the shareholders. In the 1988 season, for example, a share cost about $300 and entitled a shareholder to one large bag of fresh vegetables and fruits each week for the eight months of the growing season, and one large bag a month in the winter — potatoes, beets, carrots, and so forth. Over the long haul of the season, a share yielded enough vegetables for two people who are vegetarians, or for a family of four with a more widely based diet.

Nuts and Bolts

Up until 1989, the CSA garden was an unincorporated association managed by its co-workers on behalf of the shareholders. The main administrative vehicle was a weekly meeting of the core group that shareholders were welcome to attend. Rather than a hierarchy with one person on top giving orders, the group has striven for consensus in decision making.

"Once we had the core group," CSA member Andrew Lorand has commented, "we were on the way. The important thing for anyone starting out like this is to get a core group of four or five people who are committed to making it happen. Later, when we sent out the prospectus for the garden in the spring, it was actually quite straightforward and easy."

Andrew says community supported agriculture (CSA) is a simple concept. Because of economics and politics it usually isn't possible to do the right thing in the garden. But he says the CSA Garden takes as its starting point, doing the right thing in the garden. "We have a responsibility towards the way that we treat the earth, and only by treating the earth in as comprehensively good a way as we know how, can we expect it to properly nourish us. If you take that as a starting point, then that's something people can connect with. It's a practical ideal. For that to be possible, though, there has to be a community that says 'we want it.' Because if you do it the other way around — 'I'm going to have an organic garden because I can get a higher price for my produce' — forget it. It will never happen. There's nobody to take the responsibility. So CSA basically says, if you want to take control of your own food situation, then join us and support the gardeners. Then the gardeners will support you with clean food."

Each year over the winter, the core group draws up the budget and calculates the price of a share by dividing the costs of producing the harvest by the number of shares the garden can reasonably be expected to provide. In practice, the budget, which reflects the experience of the previous year, is the basis for the estimated price of a share. However, if they underestimate, they can charge up to 10% extra. Should there be a financial surplus, which the group concedes is unlikely, it would be refunded to the shareholders.

In the spring, the core group issues a prospectus for the coming season, showing the shareholders what the costs will be, and what they can expect in return, including a chart showing approximately when particular crops will be harvested and made available. The prospectus states, "our gardeners plant to provide a balanced harvest each week. Since there are so many variables in nature, surpluses and deficits of particular crops are likely."

The farm harvests from May until mid-October. For the first three years, subscribers had the option of picking up their shares either at the garden, or at the Berkshire Coop Market in Great Barrington.

In 1989, though, the CSA decided to experiment with home delivery. They reasoned that since people were just coming to the farm, picking up their vegetables and then leaving, the producer's group might just as well send a truck around to deliver the food —it would save energy and cause less pollution in the long run.

The development of the CSA Garden over the first three years has been solid. The vegetables have been of high quality, the amounts have been, for the most part, more than adequate, and the distribution has worked satisfactorily.

In the beginning, the CSA Garden intended that each shareholder would do two days of work each year in the garden. But that just didn't work out. People were supportive of the CSA concept, but they just didn't get directly involved. The group found it was up against many strong cultural currents — the lure of the electronic cocoon. This reality prompted much long discussion within the core group around the question of what they were actually doing: creating a community of consumers, or simply marketing a somewhat unconventional product.

By creating their own market, the CSA Garden has escaped the bind of the vegetable market in which farmers must often sell their products for less than it costs them to produce it.

An Open Secret

One noteworthy facet of the CSA Garden is the involvement of mentally handicapped people from the nearby Berkshire Village. They participate not just as consumers, but also as workers in the garden.

John Root, Jr., who was active in forming the CSA, has also been a leading force in the involvement of handicapped people with the CSA. He says his basic interest is in community, and working with handicapped people. "When you are working with handicapped people you are creating, whether you want to or not, a community around them, a community in which they can live. The handicapped person tells you something about what it is to be human, because we usually think in terms of 'I'm intelligent, or I have particular talents.' But the handicapped person doesn't have that. In creating a community in which a mentally handicapped person can live a decent dignified life, you also discover that that's the way everyone should live. This is an open secret in which everybody is gradually discovered."

"My particular interest has been in creating situations where everyone in the community is responding to the needs and the possibilities, and where people are not being coerced. As I see it, if someone hires you to do a job the salary is a coercion. You agree to do the job for a salary; now maintaining the job by doing things you don't necessarily think are right or believe in, that's all related to the fact that you are getting paid a salary. One of the things, for instance, that a mentally handicapped person can't do is to support himself with a salary, a wage. This raises all kinds of questions."

Principles and Pressures

The members of the Great Barrington CSA started out with a long list of ideals, most of which they have held to fiercely despite ups and downs, the inevitable conflicts among people, and the constraints of the modern world. Among those ideals are respect for the workers as well as for the earth, plants, animals, environmental limits, and cycles of nature. The CSA has also made a commitment to use organic and biodynamic methods, to be energy conscious, to maintain decent working conditions, to emphasize the therapeutic value of agricultural work, and to support community control of land.

What has been a problem, as even their own newsletter states, are philosophical differences among the members of the core group. The problems have arisen in three areas: the relative importance of biodynamic gardening techniques versus standard organic techniques; determining who has how much decision-making authority; and determining the role of the shareholders in the CSA. Such issues are likely to arise in any group enterprise.

The CSA Garden has also faced the challenge of land economics. According to John Root, Jr., "the single biggest issue in the development of CSAs is acquiring land. So far all CSAs either have a donor who has given them the land, or someone who is very sympathetic who is practically giving them the land, or they have a very tenuous agreement about the land . . . In the long term, we need to create a consciousness that land for agriculture needs to be taken out of the normal economic equation. We need a CSA land trust — angels who will buy land for the CSA. But one way or the other, the agricultural land must be taken out of the marketplace."

For the CSA Garden, an hour of reckoning with the question of land came at the end of the 1988 growing season. Their three-year lease with the Indian Line Farm expired and the core group faced the questions of whether to buy the farm, or move elsewhere, and whether to form a corporation of some kind to do this. Since they make decisions based on consensus, this was not an easy process.

"We knew in the very beginning that we needed to have a reasonably secure garden site. We felt reasonably happy about Indian Line Farm," John Root Jr. says, "because we had a 3-year lease with an option to buy. But we did not succeed in buying the farm because we couldn't agree on price." Ultimately, after much hard work and discussion around these issues, the CSA Garden arrived at a plan for the long-term future. Specifically, the members committed themselves to establishing a comprehensive farm that would support its community for years to come, an agricultural enterprise farm that would provide quality food for consumers, and a healthy livelihood for farmers.

The CSA Garden group intends to form a land trust which will acquire suitable land either through a 99-year lease, or ownership under a land trust arrangement. The group considers that land prices are inflated due to speculation, and they want to take some suitable agricultural land out of the economic rat race.

The group envisions a full biodynamic farm of 100 to 200 acres serving the complete nutritional needs of 200 to 225 families with milk, butter, yogurt, cheese, eggs, chicken, lamb, and beef, as well as fruits and vegetables.

As Andrew Lorand puts it, "we're convinced that in the long run the care of the land as well as the ability to meet the total nutritional needs of the consumer demands a whole farm. And the ideal, of course, of biodynamics, is to have a whole organism which takes care of its own needs . . . The other thing that we are striving towards, which is perhaps a bit different that some of the other CSA's, is being able to meet the total nutritional needs of the members of the community."

A Plan for the Future

To help realize their larger vision, in early 1989 the CSA Garden reorganized internally, forming two distinct groups: the producers and the consumers. In this way, they hoped to activate and maintain a supportive dialogue between the two. "You need two clearly defined groups to have a dialogue," Andrew comments. "Otherwise you have a monologue among many."

The producers will carry the farming initiative while the consumers will take some of the risk of farming from the producers. The producers association will own the means of production, and also the animals, buildings, machines, seeds, and so forth. And then, as they settle onto whatever land they acquire, the plan calls for the producers to also own living quarters, gaining equity over time.

The consumer's group has elected to incorporate as a non-profit organization, while the producer's group will remain an unincorporated association for the time being.

To keep the garden rolling through the year of transition, the CSA Garden extended its lease with the Indian Line Farm through 1989. Meanwhile, the members of the CSA Garden have begun working with Professor Christopher Nye at Berkshire Community College. He has developed a Farm Match Program, to match young farmers in the area with people who have land but who do not want to farm it. Together they intend to locate a piece of land where they can develop the comprehensive farm of their dreams.

(Note: After this chapter was written, but before publication, the Great Barrington CSA split and became two farms, Indian Line Farm and Sunways Farm.)

EXAMPLE 3
The Brookfield Farm

It began to happen very fast. The farmland across the road from David and Claire Fortier began splintering into two-acre house lots. As if it were overnight, suburbia took root in fields which had supported crops for decades.

As they watched the houses go up, the Fortiers became deeply concerned at the loss of the farm land in their neighborhood of South Amherst, Massachusetts. Thinking back on that time, David, a retired professor, recalls "there were about 70 acres of land here which had been abandoned by the original farm around the turn of the century. The land had been apple orchard, hay fields and pasture. But it had been allowed to overgrow for about 60 years, and then for the last 20 it was rented out for growing corn, and they mined the soil. They'd just throw in some fertilizer and some corn seed, and then take the corn out. They did that year after year, and there was not much left in the soil. When we arrived here about 1963, it was still in corn. They left the field fallow for a year, hoping that would do some good. But nothing would grow in it, not even weeds. It was like a desert with sand and the topsoil all blown and washed away."

The land eventually came up for sale, but at a premium price. So it looked like the land was going to be developed. Ultimately, though, David and Claire were able to purchase the land. But what then? They were not farmers. To keep the arable acres clear, they continued to rent it to a farmer who grew corn. But they were looking for a permanent farmer who would treat the land with full respect. For a long time, without luck, the Fortiers advertised for a farmer.

While this was going on in South Amherst, Ian and Nicolette Robb were living on the other side of the Atlantic in Aberdeen, Scotland at the Camphill-Rudolf Steiner School for handicapped children. Ian was apprenticing in the garden, and Nicolette was completing her studies in curative education.

Eventually Ian enrolled at Emerson College in Surrey, England, in the biodynamic agriculture course. Then in the spring of 1981, the Fortier's son, Thomas, visited the college and pinned his parent's advertisement for a farmer on the bulletin board. Ian and Nicolette saw the ad. A short, two-month correspondence followed, and by August the Robbs had arrived to begin farming.

Hatching a Plan

"When we arrived," Ian explains, "all that was here was the Fortiers and their hopes of having someone work the land. So it was just the four of us. We felt our first task was the land, to really deal with the land and work out a plan of what we should do. We took soil tests all over and it was a horror story because of the continuous planting of corn. The land was in poor condition, hardly any organic matter, terribly low ph, and so on and so forth. So we could see that before we could grow anything here we really had to improve the land. I mean, there were no animals, nothing, just land. A small barn had been built, and there was an old tractor and a few pieces of equipment. We had a unique opportunity of pretty much starting from scratch, and we could do what we felt needed to be done."

The Robbs looked around and saw the obvious. Amherst is located at the easternmost edge of the Connecticut River Valley in Massachusetts, one of the top vegetable growing regions in the country with highly fertile soil. There are dozens of vegetable growers, and the Robbs realized there was no way they could even come close to competing with the cosmetic quality or price of their produce. Vegetable stands seem to sit at every bend in the road.

"We felt that for something to become of our endeavors, it had to be based on a fertile, healthy farm," Ian explains. "We were facing some serious economic questions. We weren't a trust or a foundation then, we were just a group of four people. David and Claire had pledged to stand behind the farm financially for a few years to get it going but we still had to face the questions. What shall we do?"

The Robbs and the Fortiers had many meetings, just the four of them to decide how to proceed. Right from the beginning they wanted to build toward a complete biodynamic farm. They had the whole farm concept: the cows, the chickens, the grain, the crops, and so forth. But they had to start from virtually nothing.

"We had been gardening biodynamically for seven years at that point, so we had some experience and lots of theory," Ian says. "We knew where our particular talents were: berries, raspberries and strawberries. It looked like we could plant those berries here. We had grown vegetables all of our working life, but with so many vegetables available in the valley we had to wonder. Was there really a place for our vegetables here? We were a little bit cautious, let's say, about suddenly making another vegetable garden here. The organic market was virtually non-existent, so we weren't kidding ourselves that suddenly people were going to start lining up for our produce."

Nicolette explains that they had to learn about the microclimate of the farm, which is in the shadow of Long Mountain, a frost pocket, which shortens their growing season. "We always started small," she says, "almost experimentally, and then went from there. As I look back on it, that was probably a wise move; we never made any huge mistakes as a result."

The Robbs started in 1982 by creating a huge garden to grow food for their own family. The garden did so well that they had a super surplus. That's just when the Bread and Circus natural foods supermarket opened, a Godsend from the Robb's perspective. The store began to change people's awareness of the importance of natural foods, which suddenly became important to people. Interest blossomed. And as it did, Bread and Circus began to purchase the surplus crops from Brookfield Farm, as much as they could get. Soon their market garden was off and running, successful and developing a reputation for high-quality fruits and vegetables. But the Robbs were not satisfied.

From Market to Community

As they settled in and developed a feel for the community, it became clear to the Robbs that many local people had no access to the farm. Nicolette says "it was, for us, a feeling that this was just not right. At this time the farm was still the Fortier Farm, and we were still a market garden selling our produce to Bread and Circus or to a distributor in New York City. It just did not sit well for either of us. We had an underlying sense that something was just fundamentally wrong. And that's what induced us to talk with

Trauger Groh of the Temple-Wilton farm, about 60 miles away in New Hampshire."

"With Trauger we shared our frustrations, our sense of what we wanted. We were just not able to practically put it into effect. He was the one who was able to help us to bring it down into the nuts and bolts of a working operation. We needed the help and the knowledge and the experience that he already had. When I look back at all the diverse elements coming together as they did, it just seems so right."

With Trauger's help, they decided that the way to create the full farm they envisioned was to link it directly with a community of people who would pledge annually to support the farm. This was the genesis of Brookfield Farm. The concept is simple but in some ways revolutionary.

Beginning in June, 1987, 51 households joined together to support the health of the land, its crops, the overall farm, and the newly formed community. They came together out of a shared understanding that agricultural problems arise from an imbalance between the agronomic, economic and social realms. They believe the earth is ailing, and that, together, with respect, they have the capacity to take responsibility for a portion of it.

One of the foundations of Brookfield Farm is the realization that, in today's world, not everyone can be on the land. The independent families who make up the farm community entrust the care of the land and crops to people skilled in the art of farming. The farmers' financial needs, as well as all other farming expenses, are covered by the monetary contributions of the farm community membership.

The families who have chosen to associate with each other as Brookfield Farm pledge support each year, providing the financial underpinning necessary for the farm's operation. In return, the farmers and farm provide, without cost, not only the vegetable and berry needs of the community, but also the possibility for community members to have a direct connection with the land, the farmers, and the food they eat.

"We were in with faith the first year," Ian says. "We formed a small core group and established a scenario. We said, well, let's start the community farm. We don't know how many members we'll get, so we'll keep the wholesale business going, so that if it doesn't work out we'll be covered. We'll go on selling to Bread and Circus and sell at a farmer's market. But we had a strong feeling that maybe we should burn our bridges. That the community farm might never take off because we didn't have much faith in it. You know, if you really had a tremendous amount of faith in it, maybe that would help. It was scary. But what we did was to say that we would go 100% to the community farm and that we would have faith that there would be the families we need. Of course, we had been talking about it and we already had interest from some families, but not the full complement we were hoping for. So we went ahead on faith."

Nicolette adds "it didn't take long really, for us to realize that you cannot do justice to both operations — a community farm and a market garden — at the same time. We knew that ultimately the market forces would win out. To succceed in the market requires extremely aggressive selling. You have to keep your focus on the market, and you could not give your attention to the community and that dimension of the operation. There's just not room for a dual enterprise."

How It Works

Each year in March pledge forms go out to the approximately 50 families who are part of Brookfield Farm. The pledge amounts are based on the estimated budget of all farm expenses for the coming year. As both Ian and Nicolette aver, the pledge does really not buy a subscription to the harvest, but rather commits a certain amount to support of the farm. In return for that the families do get produce, but they have pledged to support the whole farm organism, not just the vegetable garden. "When you choose to support the farm, it's different," Nicolette says. "If you are just paying for some new way to get vegetables, then what happens when the farm has a bad year? Do you withdraw your support? That's when the farm needs it most."

Ian tells how in 1988 they had a deficit between the cost of running the farm and the pledges they had taken in. "Community members got together," he said, "and started brainstorming, 'what can we do?' And different activities happened, fund raisers. It wasn't 'Ian and Nicolette have a problem,' it was 'our community farm has a problem.' There were tag sales, and one community member had a restaurant in town, and if a community member went there for a

62

meal then 10% of the bill went to support the farm. People saw there was a need and actively responded. We can't generate money here. We can't say well, 'we'll sell off an acre of carrots to make some extra money.' We're feeding people, and we can't make money. There's nothing to sell." Reflecting on the way the community members responded to the crisis of 1988, Ian says "it was heart-warming. It made us feel 'this is what it's all about.' We recognized that there are people out there who are as concerned about the farm as we are, as the farmers are."

People come to the farm for their produce, where it is arrayed on tables. Community members take what they feel is necessary for the week, but the member in charge of the shop assists them with information on what crops are coming in, or which crops are in short supply — information about what is going on in the garden.

The people who make up the farm community give financial support to enable the farm to function in a radically new way, and they also have an opportunity, if they so choose, to support the well being of the farm by helping out, from time to time, with chores like weeding or thinning crops. The farmers devote themselves to maintaining nature's balance, supplementing the soil's fertility, producing the highest quality crops, and building a healthy farm without being dependent on production for profit. The community members not only partake of as much or as little of the produce as they need, but they can also appreciate the intimate union between people and earth from which, in society's present structure, people are often divorced.

Through financial pledges the community commits itself to paying whatever is necessary to cover the costs of running the farm for the year. With this commitment, the possibility exists for an increased variety of produce and the ability to provide for more and more of the members' needs each year.

As Ian puts it, Brookfield Farm has "taken economics out of the center. You know: 'give me the carrots and here is the money.' That is all taken care of by the treasurer and it happens in advance so the farm is taken care of for the year. Already we have separated the idea that the produce that you get at the shop is worth some equivalent of what you would pay at the store, or something like that. No, it's because the whole farm was made possible, right down to the pig in the forest, or the cow in the shed. Because all of that is possible, there is produce on the table. So we keep on emphasizing that how you pledge on an annual basis, the guideline for that is derived by the total budget somewhat divided — and it may not be equally divided — by the number of adults in the community. And that's not an unlimited number. We can only feed so many. So the farm is really almost tailor made to the community that is supporting it."

Farming In The Forest

Brookfield Farm is situated above the Lawrence Swamp aquifer, the major water source for the town of Amherst. Out of 70 acres total, approximately 58 acres is woodland in the swamp, so it's a very sensitive area. Without chemicals of any kind — for which the town is deeply grateful — the Robbs produce an abundance of fruits, vegetables, herbs and honey.

Their hay fields, pastures, and forest land are arranged as if in concentric rings. The heart of the farm is the four-acre vegetable garden. That is surrounded by the hay land, which is a little less intense in terms of activity, and then the forest land. The Robbs are steadily moving the farm into the forest. Ian explains: "We are seeking to permeate it completely for the health of the forest, which has suffered from acid rain. Livestock will be allowed to roam in the forest, pigs and cows. They graze on some plants and trees, and trample others. They also fertilize the soil with their manure. And eventually, over the years, the forest land begins to take on a park-like quality. We feel that this is where you can have a mutually supportive relationship with the forest. No longer just using it as a supply of wood, but the farm can, through a diversity of livestock in the forest land, start to give something back."

It was with forest grazing in mind that the Robbs chose a particular breed of cow, Irish Dexters, which graze and will do well in this setting. The Dexters can be used for beef or they can become milk cows. The Robbs intend to use their cows for beef now, because they are not yet in a position to set up a dairy.

Keeping cows in the forest is not an original idea, Ian says. "Historically, the forest has been a part of the farm. It may be more of a European tradition. We have struggled with this dimension of the imbalance of the farm. A 70-acre farm with 58 acres of this kind of impenetrable jungle as the forest grew back up after having been logged. So we felt, maybe there's a gentler way of going about it with this land. Maybe the chain saw is not the answer."

A Farmland Trust

The farmland, still privately held, is in the process of being turned over to the Biodynamic Farmland Conservation Trust — a trust created to serve as a vehicle for supporting any initiatives which may arise in connection with the farm — research, education, apprentice programs, and so forth.

The trust was formed in 1988 because David and Claire Fortier wanted to ensure that the land would remain in agriculture after they pass on. The Fortiers dissolved the commercial farm corporation, Fortier Farm, and passed the title of their land to the Trust. As David explains, "a commercial farm was never what we really wanted. The only possible rewards with a commercial farm are economic, and the chances of that are slim hereabouts. We knew the farm needed the community aspect. To keep the land open requires a gesture of sacrifice from someone or some group of people who will donate land or otherwise make it possible. Land prices are just too high for it to happen otherwise."

As the Trust stabilizes, it will help support the operation of the farm in many ways, especially with long-range planning and financial support. The Robbs and Fortiers look forward to the possibility of having the trust take on more responsibility so that others can donate land, equipment, materials, as well as financial assets. As biodynamic agriculture grows, so the responsibilities of the Trust could grow. Ideally the running costs of the farm should be met through the pledges of the families who belong to it, and fund-raising activity will happen in support of the trust, for capital purchases and capital expansion programs.

Set up along a standard model, The Biodynamic Farmland Conservation Trust is just starting to find its way. The trust is a non-profit, tax-exempt organization, so the land will no longer be subject to taxation. But the trustees are not sure they will take advantage of this opportunity. They may wish to go on paying taxes, as a good faith gesture to the larger community they are part of. However, they

65

cannot afford to pay at the rate they are currently being taxed: as if the land were open to development as house lots. They hope to negotiate a rate realistically pegged to the land's value as open farmland.

One Step Further

Armand Ruby is one of the seven trustees of the Biodynamic Farmland Conservation Trust. A teacher of environmental science, one of his principal concerns is land usage.

In recent years he has studied many alternative forms of land usage and farming. But when he learned about biodynamics, he knew he'd found what he had been looking for. "The biodynamic view of the world takes the notion of Gaia — the earth as a living organism —one step further. If the modern view of Gaia is the earth as an interdependent super organism regulating the earth's waters and atmosphere and so forth, then that view is expanded in biodynamics to place the earth in its cosmic setting: a living universe. So the concept of earth, or Gaia, becomes even broader or grander, deeper in meaning: part of a living universe.

"In the biodynamic view of the world, the earth's biota is far more than just a complex set of chemical reactions. The highly mechanistic view of the world as a set of chemical reactions is incomplete. There's more to it. There's a spiritual basis to everything that we see on the earth. When we see the universe as alive, we see that life is infused with cosmic spiritual influences, and we see humans as spiritual beings incarnated in human form. It's the same with a cow or a tomato. The physical part we see is not the sum total of the cow or the tomato. There's something more. And this is also true with the soil; it's not simply the constituents we can analyze in a lab with microscopes and so forth. In some way the soil, the earth itself, is also alive.

"Supporting a biodynamic farm does something more than just produce health-giving food for people; it also reenlivens the earth, increasing its vitality. It is health-giving for the earth itself. Biodynamic agriculture is one of our best potential means for healing our relationship with the earth. We have done a lot to really damage the earth in recent centuries, and biodynamic agriculture is a way we can start to help heal that relationship. You can look at it from just an ecological level: how does it help to build soil as opposed to deplete soil; how does it help us to avoid polluting the soil and ground and surface waters? Then you can go from that ecological level to more spiritual levels as well. How does biodynamic farming help us to build our relationship with the earth as a living being?"

A graphic illustration of this respect for the land comes in a story Nicolette Robb tells. "If you go into the farmlands of this valley in the height of summer, there's not a bird singing. You can't hear one insect, the pesticides have been used so heavily. In the eight years that we have been here there's been a change in the wildlife around this farm. Birds and animals have returned in great numbers."

A Part of the World

"This is not a community that goes off into the mountains and builds walls around itself in isolation," Ian says. "Rather, it's a community within society, within an area that's just mainstream America, and yet these creative things have arisen."

"We want people to come out to the farm. We want to meet them, and to have them meet us and to see the farm." To abet member involvement, the farm has placed a large compost bin next to the parking area. That way, when community members come to pick up their produce, they can bring with them a bucket of household food scraps and dump it into the compost bin. They don't just take away, they also leave something.

At Brookfield Farm the emphasis of community support lies not so much in the direct relationship between money and produce, but more in the community's involvement in supporting the whole farm. As Nicolette puts it, "we seek to establish a common feeling based upon our interaction with people around the activities of the farm. That was the exciting thing to our first community members — not just a way of getting produce, but the concept of supporting the whole farm."

Naturally, Ian shares this feeling. "If you want to have people who are supporting an individual farm, or a group of farms that are working together, then it really is a case of supporting the farm and taking the good and bad times along with it. We hope there's good times," Ian says, "that the farm does well, and that the animals stay healthy, and that the land produces. But I get the feeling, and it would make me a little nervous if we were doing it here, that if we had a set fee for a set amount of produce, that people might wonder why their money was going for things that they don't get an immediate benefit from. They might say, 'Well, I'm just interested in vegetables, what are you talking about pledging for cows? And besides that, there weren't enough beets left when I came to the shop.' They will always be disappointed because you promised them two pounds of onions, and three pounds of lettuce every week, and what happens if you don't get that? You feel cheated. So we haven't let this happen. Of course, everybody's different and some will feel there's a direct connection between the amount of produce they receive and the pledge that they made. That's a reality. But most feel they are supporting the farm.

"Nicolette and I are not kidding ourselves that we have all the answers, but we do try, with the core group, to always bring our practical actions from the realm of the highest ideals for the land, the group, and the earth. So then you feel that you are doing something morally correct, that you are heading into the next century doing something that is really bringing the earth and humankind forward in the right way. If you really are coming from the highest principles, and you have all the other stuff that goes on, all the bookkeeping and all the finances and the nuts and bolts of the farm — they are important, they have their place, but it's the high ideals that sustain you. That's what gives us the strength to face all the changes and all the unknown factors. What if this year there's not a drop of rain? If it's a terrible year for farming, and at the end of the year there's no produce at all? Does that mean the end of the farm? Would people still be behind the farm if there wasn't much? What happens if disaster befalls us? That's why we're not just selling food, or a share in the harvest. That's why we want to involve people directly in supporting the farm.

"What can we feel secure about today? There's very little to feel absolutely secure about. The day has gone when I as a man in my 30's could look over a piece of land and feel absolutely secure for the rest of my life, and imagine that my children will be here to work on the land, that kind of generation after generation security of the land. That has already changed. I will be interested to see how Brookfield Farm develops over the years. We want to steer it in a certain direction. We're not just open to any old whim, we have ideas. But because people are involved it's not just going to be mechanical. We have to take that into consideration. We have a plan. Every good farmer ought to have a 30- or 40-year plan. You know, this is what you're working towards. In winter moments Nicolette and I can sit before the fire and we can see a park-like landscape, with animals foraging in the woodland. We can see it, but we have to realize it may never happen, or that it may happen in some other way. Who knows? We have to allow for that."

No Stock Formulas

"There's no stock formula," Nicolette says. "Ian and I have always felt a little inner reticence at being lumped under the concept of CSA because we feel it does an injustice to Brookfield Farm and its potential. There's so much more that can come from this farm. There's nothing wrong with the CSA concept. It's really exciting for the community to take initiative to control the way its food is grown. But here it came the other way. The farm needed to be cared for, and the future of the farm needed to be cared for. And it was from that vantage point that this group of people came together to embrace Brookfield Farm. There's a different emphasis.

"Some of the ideas we're working with are fine for a model," she says. "They can give you things to think about — issues of ownership and financing, and all the nuts and bolts. But we soon realized that there was no easy route. We couldn't pick up a paper describing some other farm and place it down over our land as if it were a model, as if you could stamp it out on an assembly line. And you couldn't pick up the model of our farm and place it over another 70 acres and say that it's going to go exactly the same. I doubt that."

Does the broad model they are establishing have potential for the future, or is it only for a sophisticated handful? Nicolette says "There are elements that can be duplicated. But you need to acknowledge the wider ramifications of this, that you're not just dealing with the economic structure of the farm, and its need for economic support. The relationship that can develop between the farm and its community allows for a mutual support system that encompasses far more than economics alone.

"If the farm is solely an economic model," Ian adds, "and you are attracting your support purely on an economic basis — it's cheaper or some such — then it's all in the realm of the economist, not the farmer. Because all you would need then is for an even smarter economist to come along and think of an even smarter economic plan with slightly different variations. We can't think of everything. There's probably better ways of doing this. But we are receiving the support we need."

"We're all pioneers here of this way of farming, of this kind of relationship to the land, but I'm sure we are the predecessors of the predecessors of the new type of farmer. There have to be many more farmers involved with this, and it will have to stand the tests of time."

EXAMPLE 4
The Kimberton CSA Garden

In the hilly countryside of Chester County, Pennsylvania, astride the swift waters of French Creek, the Seven Stars Farm has been a home for innovative experiments in agriculture for over 50 years.

Originally part of the A. Myrin estate, the land where the farm is located became a training center for biodynamic agriculture under the guidance of Dr. Ehrenfried Pfeiffer in the 1940's. With hundreds of people in America and abroad, Pfeiffer shared his experience as a farmer and technical advisor. Eventually, 400 acres of the farmland were donated to the Kimberton Waldorf School, which still holds title to the land. Now, on a small portion of that land, a CSA garden has taken root.

Just down the road from the school and the garden is the Bio-Dynamic Association for farmers and gardeners who are interested in this approach to working with the land. Rod Shouldice, Director of the Association, was instrumental in organizing the Kimberton CSA. He first heard about the idea from Jan Vander Tuin in 1984. Vander Tuin, who also inspired the Great Barrington CSA, was tremendously excited about the idea and wanted to make it happen immediately. But the time was not right. People just weren't sure the idea could work.

As Rod recalls, at least one organic grower experimented with subscription farming in the Pacific Northwest in the late 1970's. He created an arrangement whereby people came in the spring and placed orders for potatoes and carrots and so forth, agreeing to pay a fixed price when the harvest came in. But consumers were still buying their produce by the pound, not contributing directly to the support of the farm, as with the CSA concept, where the consumer agrees to take the risk with the farmer.

In the fall of 1986 the Seven Stars Farm was searching for something new. The school had been running the farm, and it had just not worked out well. Then Trauger Groh visited and gave a talk on the experience of the Temple-Wilton Community Farm, which had just completed its first year. The Kimberton farmers were interested, but they felt they could not just shift right over to a

71

community-supported farm. The farmers were unsure they could find the right level of support because the scale of their operation was so large: 400 acres, a scale which requires intensive capitalization to succeed. So after discussion they decided to start with a small-scale project, a five-acre vegetable garden. Then people could gain a gradual experience of what a CSA was like.

Here's the concept the community decided upon: they would form a CSA (Community Supported Agriculture) where a group of families would pledge together to cover the costs of the garden, including a decent living for professional gardeners. Then, once or twice a week, mature crops would be harvested and divided among the shareholders. They would get a regular supply of fresh, healthful produce during the growing season; and the growers would get an assured living. The farmers would not only be freed from the need to take upon themselves the financial risks inherent to farming, but also from the need to go selling in the midst of the growing season.

Shortly thereafter the Kimberton group contacted Kerry and Barbara Sullivan, who they knew were looking for a situation, and encouraged them to pack their bags. According to Rod Shouldice, this was the logical next step. "A crucial ingredient for a CSA to succeed is having gardeners who know what they are doing." In fact, one limiting factor for CSAs is finding capable people who can grow excellent produce on a large scale.

The group asked the Sullivans what it would cost for them to grow vegetables that first year. The Sullivans said $24,000 (see appendix E). Once an agreement was struck, the Sullivans also advanced the CSA $10,000 of personal savings to buy the necessary equipment for a five-acre garden. The loan is paid back a little each year.

A Late Start

Kerry and Barbara Sullivan had achieved a measure of fame as the coordinators of the North Carolina demonstration garden established by *Mother Earth News* magazine in the 1970's. But after four years they had journeyed to Emerson College in England to deepen their knowledge of biodynamic agriculture. While they were away they hoped that there would be a growing interest in biodynamic agriculture in North Carolina, and that they would be welcomed home for their expertise. But when they returned there was nothing like that. They toured several biodynamic farms without finding a situation that seemed right for them. So they returned to North Carolina for a year.

When the families of Kimberton decided to start a CSA and called the Sullivans, they didn't hesitate. They arrived ready for work in February, 1987. When they arrived, there was no greenhouse, no tools, not even a spot for the garden, just a big hayfield in contour strips. The Sullivans set to work immediately, ordering seeds, buying a tractor, and securing all the necessary equipment. In Southern Pennsylvania, farmers begin planting in March, so they had to rush.

The late start was a big handicap. Kerry and Barbara worked extra hard to build soil fertility and to fight weeds from strangling the 33 crops they had planted. But ultimately they brought in a fine harvest.

While the Sullivans were learning about the soil and the climate, the consumers also had to go through a period of adjustment. They had to learn to eat produce in season as it was harvested, instead of just picking by whim from endless choices in supermarkets.

The Mechanics

As Rod Shouldice recalls, "when we decided to start the CSA, families were willing to commit their money even though they had yet to meet the Sullivans. We walked out of our organizational meeting with checks totaling $15,000 — that's serious money. Even though it was just part of the $24,000 we needed, it was enough for us to see that we could make it."

The first year the Kimberton CSA had 60 families. Then the phone started to ring. People had to be part of the project. "Once a CSA project has worked in the world for a year," Rod points out, "you are on solid ground."

They have decided to hold at 100 families for now, so this group of 100 families has to meet the costs of the CSA for the year. As with other CSAs, each year the core group meets and makes up a budget for the coming year based on the expected expenses for the coming year. Then they call a meeting for all the families. At the meeting the families pledge financial support for the CSA. The Sullivans have found it important to have people make a deposit or to pay in full early in the year so they will have cash on hand to buy seeds and supplies.

As it stands now, one share is approximately enough for a family of four, but there's such a range in eating habits that one cannot depend on that rule of thumb. Some families of two might eat a lot of vegetables and easily consume one share, whereas a family of five or six might eat few vegetables and so find one share sufficient. Some couples use more than a family of eight. Families who do extensive canning and freezing will invest in two shares.

Not everyone pays the same amount per share. As Rod Shouldice explains, "a childless two-career couple can afford to support the garden more generously than a single-parent family, and we wanted both to be able to be members. So when we had our February membership meeting, we told people they could pay from $270 to $370 for a share, as long as the total averaged out to $320. Everyone wrote down the support they could offer, then someone gathered up the papers, went in another room and tallied it all up. It turned out we were short an average of $5 a share, so we passed the papers back out for another go-around. This time, we ended up with a surplus of $10 a share!"

In 1988, for example, a share cost each family an average of $320. Because of the drought, vegetable prices in local supermarkets skyrocketed, and the cost of an equivalent amount of vegetables was $530. So members of the Kimberton CSA not only had the benefit of fresh, chemical-free local produce, they paid $210 less for it (see Appendix E).

The Kimberton CSA wanted to create a way to share the harvest according to need. While the Temple-Wilton CSA in New Hampshire simply lets its members come and take what they want, that didn't feel right to the folks in Kimberton. Some people were worried that the early birds would get the best of the harvest. So the Kimberton CSA took a middle approach. A sign in the distribution shed on pickup day tells members how much of each crop they are entitled to take. If you don't want all of your share, you simply place it on a surplus table. Anybody can take what they want from that.

Beyond saving money, which may or may not happen, why else would anyone want to join a CSA? Barbara Sullivan responds by asking "have you been buying your vegetables from the market? Anyone who buys supermarket vegetables would have two darn good reasons to join a CSA: quality and freshness." Kerry adds that lots of people join after tasting a sample: "they visit neighbors who are members of the CSA. After they've had dinner and taste the food, there is no comparison. They have a hard time going back to the supermarket, where the food has very little flavor. Also, people know they are supporting something environmentally sound. Consumers support the CSA because they know that this approach to farming helps to heal the earth. And, they sense the difference in nutritional value. That the food had more vitality and life to it."

People also join because of the community impulse. It is a group of people doing something good, something aimed at a healthy future. As with many other CSAs to date, there is a waiting list of people who would like to join.

Pluses And Minuses

As they have worked within the CSA concept, the community in Kimberton has identified several distinct advantages and disadvantages.

According to Barbara, one big advantage is not having to market during the growing season. They put a significant amount of time into communicating with all the families and so forth, but they do it in the winter. They don't have to go out and sell while they are growing.

Another aspect that appeals to Barbara is that there is little waste. "You don't have to toss things away because they are not all a uniform size. But uniformity has nothing to do with quality. Our produce is generally the same size, but not carried to an extreme where every cucumber is exactly seven or eight inches long, and the rest are thrown away. Our produce looks great. It's all harvested and washed and displayed beautifully in the area where people come to pick up. But there's no rigid uniformity, and very little waste."

A third advantage is that the consumers know where the food came from, and the growers know where its going, and who is eating it. They talk with each other. Consumers tell the growers whether something is good or not, and they can complain directly to the growers if something is wrong. "When we were deciding what we wanted to do," Barbara says, "one thing I knew I didn't want to do was to be a market gardener and just send the food off somewhere.

I wanted to feed people that I knew. I wanted to be more self-contained. It was and is important to me to have a direct connection with the people who eat the food."

Kerry also has strong feelings about this direct connection between consumers and growers. "People will tell us, 'you know I could never get little Johnny to eat beets, but now that he knows they are from the garden, he's happy to eat them.' Little things like that are important to us."

Barbara says "with the CSA concept people don't just look at their families and try to get the best deal for their family. They look at a community of 100 or so and try to do right by the whole community. We are all trying to eat from the same garden. We don't all make the same amount of money, but then we don't all pay the same thing for our share in the garden. One person will pay more out of their free will because they know they have the ability to do so, and another person will pay less because they don't have it."

Kerry and Barbara agree that knowing they have a set income is a real advantage. The consumers take the risk with the growers, even if the crop fails completely. "Last year we couldn't irrigate our potatoes," Kerry explains, "so they failed in the drought. We didn't have enough equipment to water everything, so we just had to let them go. We had to make a decision. But the whole CSA community takes the risk and so takes the loss. We still had enough income to go on with our lives without getting into debt."

While the Sullivans may not have gone into debt, their income is low. In 1987 Kerry and Barbara earned about $16,000 (not including $2,700 from the CSA in repayment of a loan). While that's a low wage for two people who each put in about 55 hours a week of hard work, they reason that the CSA is in its infancy and that like any commercial enterprise it may take a while to begin paying off. Also, they are well aware that if they had become market gardeners they would likely have worked harder — spending many hours selling the produce — and most likely would have ended up with less income. And for all that, they would not have had the satisfaction of being involved with a new concept that reknits the community together and is gentle on the earth.

Another disadvantage according to Kerry is that it's fairly hard to mechanize to the extent a big commercial operation would, because you have so many different vegetables. "Instead of being a market garden where you choose four or five crops that you can do really well, you have to grow many different things. We grow over 50 different things — so its like a huge family garden, rather than a market garden where you have a half acre of peppers, and an acre of cantaloupes. But Barbara adds that while this is a disadvantage from one point of view, its an advantage from another. "It's what we were looking for. We'd be bored with just five crops. And also we are interested in training other people. This gives us all much more to study and learn from, a much better training situation."

Exploring Biodynamics

The Sullivans began to learn about gardening in-depth when they apprenticed in California with Alan Chadwick, the man who brought French Intensive gardening to America. Then they went to North Carolina in 1978 and founded a demonstration garden for Mother Earth News. The magazine was starting EcoVillage, a 600-acre site demonstrating some of the ideas explored in the pages of the magazine: solar architecture, organic gardening, and so forth. While in North Carolina, the Sullivans lived in a Yurt, gave demonstrations and workshops, and took on apprentices. From Alan Chadwick they had learned much, but it was at Mother Earth News garden that the Sullivans really became interested in biodynamic agriculture.

With advice and encouragement from Peter Escher, a biodynamic consultant, they began to experiment with some of the biodynamic practices. Then in 1983 they went to Emerson College in England to study biodynamics in depth. Why go to England? According to Barbara, "Emerson College was the only English-speaking place we knew of that offered a strong theoretical grounding in biodynamics as well as practical training.

"When we went to Emerson we didn't necessarily learn a bunch of facts and formulas," Barbara says. "It's more like learning a new way of thinking so you can solve your own problems. We don't just go to a text book and open it up and see what a solution might be. After all, nature just doesn't work that way. You can't apply the same solution twice in the same way. It doesn't work.

"At Emerson we did not learn the chemical kind of thinking

where you take this much out of the soil and you have to put that much back in, so-called quantitative or mechanistic thinking. But what we learned is much broader. It takes in factors other than the chemical processes that go on. It's an approach that tries to understand the life of the soil. The soil is not just dead minerals. That's important to be aware of if you are working with the soil to coax life out of it."

"We didn't go to Emerson because we had heard ideas and thought, hey, that's a great idea. We went because we tried the biodynamic preparations and approaches in the garden and they worked. So we went for very practical reasons. We saw the results first and we wanted to know why, why does this work?"

Kerry offers an example of this from their direct experience. "We could see the difference in making compost with the biodynamic preparations, and making it without. At that time we were working on a very intensive scale in North Carolina, where we could see results very easily. Here, where we have ten acres to take care of, we can't look at things as closely. But when you work a half-acre with a spade and fork you see a lot more. So we noticed things using preparations in the greenhouse, in the compost, and on the plants. It made a big difference. The compost broke down a lot faster. But what we were mainly impressed with was how thoroughly it broke down, and how much better a product it was in the end. It smelled a lot better. In the greenhouse, seedlings rooted much more thickly. And it was just better all around."

Rod Shouldice, the Director of the Bio-Dynamic Association, adds that "rather than just trying to feed plants, biodynamics tries to make the life in the earth stronger and more vital. The earth is made sensitive so it can give the plants what they need. It's a crucial difference. The day of artificial fertilizers and pesticides is over. It's dying now."

Gardening Techniques

The Sullivans employ a number of innovative techniques to grow vegetables for 100 families. They prepare, fertilize, shape, plant, and weed hundreds of permanent raised beds mechanically, using methods they learned from biodynamic grower Mac Mead in Spring Valley, NY.

They have a 1950 Farmall Super-A tractor with well-spaced front wheels so it can straddle their four-foot wide growing beds. To make the planting beds, they hook up a disc harrow behind the tractor and then later make passes with a specially built bedmaker. The bedmaker, designed by W.W. Manufacturing Co. (60 Rosenhayn Ave., Bridgeton, NJ 08302) has one large disk on each side to push dirt to the middle, adjustable boards to shape and level the soil, and two rows of small, soil-pulverizing disks (called Meeker harrows) to do the final texturizing. All they need to do is to make a couple of passes with this rig and a bed is worked up for planting.

Then they pull a homemade row marker down the bed, and it makes three precise planting furrows. Finally, they push a Swedish Nibex seeder down the rows by hand. The seeder drops accurately spaced seeds in place. Seedlings, though, are planted by hand.

When it comes to weeds, the Sullivans use a special-ordered Buddingh Wheel Hoe (Buddingh Weeder Co., 7015 Hammand, Dutton, MI 49316). The hoe has a double set of gear-spun wire baskets, and it fits the row spacings in their beds exactly. When they pull it down the rows, the baskets churn through the soil and uproot the weeds. Any weeds not reached by this contraption are removed with hand-held scuffle hoes.

In the greenhouse, the Sullivans make use of Speedling seed-starter trays, while out in the garden they use drip irrigation to conserve water, and floating row covers to protect tender young plants from frost and insects.

Seven Stars Farm

The Kimberton CSA is now nestled into a corner of the 400-acre Seven Stars Farm, but the garden and the farm are not completely distinct organisms.

For a long time the Kimberton Waldorf School tried to run the farm by hiring farmers. But that arrangement was generally unsatisfactory. So in 1988 they decided to lease the farm to David and Edie Griffiths, who agreed to farm it biodynamically.

In 1989, with the other people who help farm the land, David and Edie Griffiths began the process of making a long-term commitment to the land, entering into a 29-year lease with the school. A committee from the school will oversee the lease with the farm. The farmers will oversee the land and maintain the buildings.

The Griffiths plan to form a cooperative corporation of all the people who work on the farm or in the farm store, and then to assign the lease to the corporation, modeled after the Spanish Mondragon system (See Appendix B). This, David says, flies in the face of many of Trauger Groh's ideas, because it concentrates all the responsibility on the farmers and workers, rather than spreading the risk throughout the community. But after much deliberation, the Seven Stars farmers feel this new form is a good thing, and that it will work well for them.

With this cooperative approach, the management of the farm is creating a corporation that will live on beyond the individual people who are involved. The leasing arrangement will solve the problem for them of binding up a huge amount of equity in a major banking risk.

The school, meanwhile, is attempting to sell the development rights to the land to the state under a new Pennsylvania law. That will ensure that the land is always used for agricultural purposes, a real concern as development continues to spread out from Philadelphia.

As for the CSA garden and how it fits in, David Griffiths describes it as a symbiotic relationship. The CSA makes an agreement with Seven Stars Farm to lease the acreage for the garden — five acres in 1988, ten acres in 1989. Beyond that, the particulars of the relationship are still in the process of being worked out.

David is an enthusiastic supporter of the CSA concept. "I must stress the importance of a CSA for a large-scale farm operation. Everyone talks about farm preservation, but this is something concrete. It brings people into direct relation with the farm and the farmer when they come to get a box of food. There's real contact with the farm, and that builds support for the farm as a whole. It's something real, not an abstraction you read about in the paper."

Sales at the Seven Stars Farm Store have jumped nearly 25% since the CSA began. When people come out to pick up their vegetables, they naturally stop into the store to pick up other supplies. So the CSA can help anchor a large, non-diversified farm by forming a support network.

Both the CSA and the farm are doing the best they can separately. So at the moment joining together in some more deliberate way seems to be an unnecessary complication. But that may happen later down the road. The Sullivans get everything the farm gets for the same price the farm pays. They use tractors or other heavy equipment for an hourly fee, and pay a small sum for manure and compost from the farm. It's a busines arrangement. But as David points out, "we can't let business hide the fact that we are all farming the same farm. It's not our manure, or their manure. It's the farm's manure."

Realistic Idealism

Kerry and Barbara Sullivan intend to stay with the Kimberton CSA over the long term. Barbara says "in some ways our scope is more limited, because we are not trying to create a whole farm organism within the CSA, because we are part of a dairy farm and that's where we get our manure. We don't need to get animals. It's a real difference from an operation where the CSA is the whole farm. Our CSA is just the vegetable garden within this farm. That makes it much easier on us."

The Sullivans have no overwhelming urge to expand. As they see it, there's a whole world in seven acres — lots of room for refinement of what they do, and improvement in the CSA. "Within what we are doing now, we could take a major step to growing most of our own seeds. That's a big step we aren't ready for just yet," Barbara says.

The CSA concept is catching on, but as each group discovers, there's no formula. Because the people are different and the circumstances they face are varied, each CSA does it its own way. As Rod Shouldice says, "Every project is different, and should be different. One CSA can't be reproduced elsewhere. The group that's forming it must be realistic about where they are at. I encourage them to stretch their idealism — but still they must be realistic."

Rod believes the CSA movement will spread widely. "I sense a continued strong interest. People frequently call the Bio-Dynamic Association for advice. But the CSA is not a finished product. It's not perfect, but at least we feel we are heading the right way. If we can resolve the problems, then CSAs are definitely here to stay.

"It's easy to imagine vacant city lots bursting with organic vegetables, or small groups of city dwellers getting together to hire a gardener to produce vegetables in the suburbs for them. It could even work for a corporation to hire gardeners for a six-acre plot behind the corporate center, and then saying to employees, 'Besides picking up your pay check once a week, we're going to give you a basket of vegetables, too.' "

The media is very responsive to the idea of community supported agriculture, according to Rod, so a core group can take advantage of this to help them get started. He recommends first having a public lecture to get people aware of the idea, then a follow-up meeting, and possibly a story in the local newspaper or radio station. "By then," he says, "you'll know if there are enough families willing to put their money on the line."

EXAMPLE 5
The Hawthorne Valley Farm

A rangy man of medium height, Christoph Meir has a farmer's strength in his arms and in his convictions. He takes his rest gratefully, for the moments of repose are few and far between. Since 1974 Christoph has been director of the Hawthorne Valley Farm, an innovative enterprise located in Columbia County, about 100 miles up the Hudson River Valley from New York City.

The Hawthorne Valley Farm embraces 350 acres of rolling fields and forest, land rich with streams and ponds. This beautiful valley is the stage upon which a sizeable community of people have joined forces to integrate agriculture not only with education but also with art. Hawthorne Valley stands alone, different from the typical community supported agriculture (CSA) concept. The farm operates under the umbrella of the Rudolf Steiner Educational and Farming Association, a non-profit, tax-exempt organization, which also includes a school, a visiting students program, and a number of artists and artisans.

As stated in the Association's brochures, Hawthorne Valley has a broad vision: "The age in which we live demands new capacities in all human endeavors: in the sciences, arts, humanities. These capacities are insight, purpose, practicality — the necessary foundation for action and leadership. At Hawthorne Valley the interaction of education, agriculture and the arts is intended to provide the basis on which such new abilities may be fostered." Since its founding, the Association has made notable strides toward realizing its vision.

How the Farm Came to Be

In the early 1970's Hawthorne Valley was primarily a visiting farm for educating young people about agriculture, giving them first-hand exposure to life and work on a farm. Then, as now, students and teachers worked together during a week-long stay, sharing the daily farm chores, seasonal projects, hiking, crafts and nature studies.

But with this limited focus the operation was not prospering. The people involved found it difficult to run both a school and a farm, so they sought a farmer to take full responsibility for the land. Ultimately, they found Christoph Meir.

Christoph was farming near Geneva, Switzerland when he heard about the opportunity at Hawthorne Valley. The association had the farmland and the buildings, a few pieces of equipment, and $40,000 to fund the first year of operation. With this skimpy purse, Christoph had to buy the seeds, the cattle, the equipment, to hire help, and to pay his own salary over the first year.

As he began, Christoph insisted that the farm be primarily just that, a farm, not a school farm. While students would be welcome, the main business of the farm would have to be farming. The visiting programs would have to be run by others whose main business is educating young people. When that agreement had been established, Christoph joined with other workers and began to build what is today a modern, full-scale diversified farm operation.

Virtually self-sufficient, the Hawthorne Valley Farm relies on its own feeds, manure, and compost. No chemical fertilizers, pesticides, or herbicides are used. Instead the farmers follow biodynamic principles to produce milk, eggs, grain, vegetables, meat, and honey.

According to Christoph, a typical year's planting includes 30 acres of wheat and rye, 30 acres of oats, peas, and barley, 30 acres of corn, and 110 acres of mixed grass and clover to feed the cows, horses, pigs, geese and laying hens. Five acres of the farm is planted intensively with vegetables.

In 1988 the farmers completed building a new dairy processing plant, where they produce cheese and yogurt from the milk of their 75 cows. The farm also has a bakery, which specializes in breads baked exclusively with the farm's own stone-milled wheat and rye flours. Most of what is grown and processed on the farm is sold directly through a farm store which has become so highly regarded that people drive miles out of their way to shop there.

The farm's gross sales are about $750,000 a year, which includes about $270,000 from what is produced and processed right on the farm. The rest comes in from the retail store and the other dimensions of the diversified farm operation.

Alternative Business

Gary Lamb is the Associate Director of the farm, and the manager

84

of the farm store. He says that when he first came the farmers were selling milk, cheese, and bread. They had a little door on the edge of the barn where people put their money in a cigar box and picked up what they wanted. But after business picked up, the farmers got tired of having to deal with the cash box and supplying the milk and everything, so they decided to hire someone to take care of it. But to justify paying someone, it became necessary to add other products from other farms. One thing led to another, and the Hawthorne Valley farm store developed from a cigar box in a corner of the barn to a modern retail store with five employees and $400,000 in sales each year. As Gary Lamb sees it, "the store has grown from supporting itself and the farm, to supporting alternative agriculture in general."

The farm store sells bread, cheese, yogurt, and a wide variety of health foods, fresh produce, cosmetics, books, and crafts. The store also sells the farm's products by mail order and at markets in New York City.

Community Support

Hawthorne Valley has come a long way since the 1970's. But while the farm is part of a larger association of a school, artisans, and the staff of the visiting students program, it has not yet found the perfect mix.

As Christoph explains it, "community support is our weakest part. Before we got the retail store we had kind of a food coop, which was outside of the farm, and also a little farm stand where people helped themselves and just left money in a box. We tried to have a farm producers-consumers association, and we had some meetings to discuss prices, but we realized that we farmers had to organize all the meetings. If we didn't push it, it didn't happen. So, we talked and said 'if the only interest of the consumers is to get cheap products, then forget it. We are busting our backs and not making ends meet.' So, we went more toward the retail. That was in 1979 and it marked a new beginning here."

"We felt there was no community support at all, except for some of the older friends who had struggled with us in the beginning. And we felt we were being taken for granted. People were grumbling, 'why is our food so expensive? Why does it cost more than the supermarket?' A lot of people either had their own food coops going or they just went to the supermarket — members of our community, people who had children in our school, would do a little token shopping here. It came so far that we went to a meeting and told people that if they did not change, we would have to give up. We send our children to the Waldorf School at the farm and make a big sacrifice; we pay for that school out of our farming. It costs the farm about $20,000 a year. That's a considerable financial burden. Most farms don't have that kind of expense." While this is true, most farms pay taxes; as part of a tax-exempt organization, the Hawthorne Valley Farm does not.

Finally, Christoph explains, the farmers went to a meeting in 1987 where they announced that since they could no longer afford to send their children to the community Waldorf school, they were going to start their own school. In the intense discussions which followed, Christoph says, the whole community realized that the farm, as well as the school, required their wholehearted economic and moral support.

An Associative Economy

One of the strongest features of the farm is the economic association among farmers, processors, retailers, and eventually consumers — though the consumers haven't yet been fully involved because they are not organized.

Christoph explains the farm's concept of association by referring to the work of Austrian philosopher Rudolf Steiner. "He gave an alternative to communism and capitalism. He indicated that what you have to do is that you must consciously separate an economic sector such as producer, processor, or distributor, and that they should be organized on associative principles . . . Hawthorne Valley is trying to apply certain of these principles, as far as we can — but we can't do it all in the existing social-cultural context. We can do it small, as sort of a model.

"Early on we realized that we have to develop associations where we discuss prices, where we discuss the needs between farmers, processors, and retailers. You just realize that as a farmer you are always at the short end of the stick. You are depending on what they are willing to give you. The retailer says 'well, I need so much for my living. I have to have so much of a markup,' and the farmers just cannot do that. So we decided to integrate the various facets and become our own processor and retailer. So we have these seven businesses together under one roof: a farm operation, a milk-processing plant, a bakery, a retail store, farmer's markets in the city, a mail-order operation, and a tractor dealership that sells about 12 tractors a year.

"Now the prices I need as a farmer for my milk, not speaking now as the director of the whole operation, but just as the farmer, I need $14 per hundred weight, otherwise I cannot make it. And the cheesemaker will say 'yes, but if I have to pay so much for my milk, I really have to charge more for my cheese and my yogurt.' And so we talk again with the storekeeper or wholesaler, who is tied together with us under the association, and he has to sit and listen and we have then to understand and agree with one another while the other one meets that margin or price, and whatever quality questions there are. And we have to come to consensus.

"The only thing we are missing, really, is the consumer representative of our association, who would then sit together with us and explain the consumer's point of view and tell us why prices are good or bad, or comment on the quality. The customer and the producer then can negotiate and maybe come to a compromise."

The consumers directly support the farm, but in a free way. For example, they help to finance the farm whenever capital is needed for development or expansions. A few years back when Hawthorne Valley wanted to build the farm store, they sent out a letter requesting loans. They offered what the money market was offering at the time, about 8% interest. They were able to raise nearly $300,000, and to get the money for a much lower cost than they would have to pay to a bank; meanwhile, the community members were still getting a reasonable return on their investments. Additionally, they had the satisfaction of knowing exactly where their money was and what it was doing. When the farm does borrow from community members, the loans are for three or five years, and lenders have the right to recall their loans with just three months notice. The farm issues quarterly statements so everyone can see how they are doing with their business and in repaying the loans, and the books are open so any community member can come in and see what is going on. All of this helps to create a strong bond between the farmers and the consumers.

The Farm Workers

The Hawthorne Valley Farm provides enough for its 15 farmworkers to make a living: about $1,000 a month in pay, health insurance, living accommodations, and tuition for their children at the Waldorf School, a top-notch private school that otherwise would cost about $3,000 per child each year.

Yet the workers build no equity in the farm or the houses where they reside. Technically, they are all employees. According to Christoph, they don't want any ownership or equity in the farm. "I strongly feel," he says, "based on Rudolf Steiner's indications on the three-fold social order, that farmers should have access to the means of production to use as we see fit as long as we can work them. If we don't produce, we should pass them along to somebody else who is able to use them."

But as it stands now, when workers at the Hawthorne Valley Farm retire they have no pension or other means of support. This is one of the many issues that the farmworkers are grappling with now, to find a way to provide for their later years. At present the economics of the farm make a pension program impossible. Christoph feels the only sensible way to meet this need is to improve the economy of the farm: "You should be paid enough for your products so that you can make a decent wage and so that you can also provide for your old age. It shouldn't be any different than any other business. We have raised four children here. We might not have been able to put any money aside for our old age, but our children have been given private school educations."

Christoph says the farm would also like to involve more apprentices. To date, in fact, the development of the farm has been possible only because of the dedication of the apprentices, who earn even less than the workers but who are gradually trained in the sophisticated techniques of biodynamic farming. Hawthorne Valley intends to develop a comprehensive training program, including classroom work in the winter, for about 20 apprentices.

The Question of Land

While at Hawthorne Valley the farmers do not have ownership;

the Rudolf Steiner Educational and Farming Association has granted them the right to work with the land and to use the products of the land as they see fit for as long as they are capable of doing it. Thus, at the end of their work, the farmers will not be able to sell the land for their own gain. The Association holds title to the land and the property for the benefit of the community and to ensure that it is used productively and preserved for agriculture.

Just knowing that the well-intentioned Rudolf Steiner Association owns the land is insufficient security for the farmers, so they have approached the American Farmland Trust. With the Association, they intend to deed the development rights for the farm to the Trust so that the land can never be subdivided or sold for other purposes. That way if the Association were ever to go bankrupt and the land to come onto the market, there's a possibility that another farmer could buy the farm and continue farming here. Real estate values are so high now — about $10,000 an acre in the region — that nobody could ever farm the land at that rate.

Christoph feels strongly that in general, for all farms, land must cease to be a commodity. Either the farm must give away or sell the development rights, or find some other way to remove the land from the pressures of the open market. "Today farmers have to deal with buying land and making mortgage payments, but that should not be a concern of the farmer. Land for the farms of the future will have to be provided by the community at large, or some other way, to relieve the farmer of the burden of paying for land in a situation that's really not economical. Land prices have no relation to reality or incomes, and the farmer is the one who feels this the most. That problem must be solved."

In contrast to private ownership, Hawthorne Valley's land is held in trust for the community so that no individual really has direct ownership of the land. The community supervises to see that those who are using the land are using it appropriately.

Healthy Experiments

Reflecting on the experience of his own farm and the growing CSA movement in general, Christoph says "I consider what is happening at the Temple-Wilton farm to be a healthy experiment.

It's important for a community of people to be working together to understand again what farming is about. But I don't think it is the wave of the future. I think instead we are pushing into an associative economy, which is not going out to make as much money as you can, each one for himself, but rather to understand why we have to provide each other with the means of income, so I can produce for you and you can produce for me."

Christoph regards the experimental Temple-Wilton Community Farm as a form which can work in certain situations. But, he adds, "I don't think that it's a solution for a large population. Temple-Wilton is very important, but for us it's not the right way. We want to push more into something which can lead to widespread use —an associative economy, which means, in a way, price fixing, excluding competition. The producer will tell the consumer, 'this is what I need to produce a bushel of wheat,' and not a competition to see who can make the cheapest wheat on the market, and that will be the one who makes it while the other farmer goes bankrupt."

"What's wrong with setting it up so the one who survives is the one who can produce the cheapest wheat? That principle can work in manufacturing or with industrial goods," Christoph says, "but what happens when you apply it to a living situation like health and hospital care? You can provide cheaper health care by sending a robot around to the beds to make them up and bring tea to the patients. Yes, it's much more efficient. You can build a big hospital, and you can have a computer screen where the doctor sits and decides who needs what. But if you apply the same industrial-economic thinking which is fine for building cars, if you apply that to life, to growing food, to working with nature, or to health care or education, it's not going to work. It brings destruction."

"So the fellow who provides cheap eggs, what is he going to do? What has he done in the past? One fellow figured out that if he controls the environment of the chickens — lights, temperature, air, puts five or six in a little cage on a conveyor belt — it's very efficient. You can produce eggs much cheaper. But what are we doing to the animals? What is the quality of the product? It's cheap, but is it really cheap to eat those eggs? You haven't figured in your health costs, all the diseases and mental disabilities which you might get from eating crappy food. It costs in the long run. You have to include those costs also."

"What we need, really, is a different way of thinking, a different approach. You see, our big problem is not farming. We know how to farm organically and biodynamically. We can improve a lot there, but our problem is the social and economic structure in the world. Because if we continue this way we are going to destroy our forests, our waters, the ozone layer, and our children with our school systems. We destroy our social relationships through the health industry. I mean, just look and listen to the words applied to it. It's incredible. I mean how can you have a 'health industry' that's considered an 'investment opportunity' for people to make large profits from illnesses? We need new thinking here."

With conviction, Christoph says experiments like the Hawthorne Valley Farm and the Temple-Wilton Community Farm are essential specifically because they help to test new thinking and new models.

The Need for New Thinking

Christoph raises many provocative questions about farm economics. "What are you forcing your fellow man to do when you make economic choices at the market? Are you forcing him to spray his land with herbicides and pesticides because he cannot afford to work organically, because you are not willing to pay for it? Do you force your fellow man down in Brazil or Nicaragua to spray and chop down his woods and plant coffee for you because you're not willing to pay the price of clean food and coffee?"

"We are seeing a change now," he says. "Big businesses are jumping into organics. They realize that consumer perceptions are changing about poisons in the food. So Kraft, Nabisco, and all the rest are going to move into it. But if you don't change your thinking, and you're still into the corporate way of money making and profit maximizing and industrial efficiency, then you really change nothing.

"You're going to grow food a little bit healthier, but you're still not talking about self-sufficient farm organisms which can produce their own fertilizer. You're actually only adjusting your big monocultural approach to organic — starting to import all kinds of organic fertilizers — your way of thinking and your procedures are not going to change.

"If we integrated farms with a vegetable operation where we grow 30 different kinds of vegetables for a small market, we can never compete with the fellow who grows 30 acres of carrots only, and he specializes in that. He has one kind of equipment, one setup." Why is that a bad thing? "Because," Christoph says, "such a farm has no diversity. It relies on monoculture with no crop rotation. The farm will always have pest problems which the farmer will have to deal with using pesticides, whether they are synthetic or organic."

The Need for a Broader Context

Now the Assistant Director of the farm and manager of the retail store, Gary Lamb, came to Hawthorne Valley in 1985. He was looking for a business position, and he wanted his children to have an opportunity to attend a Waldorf school. He was also concerned about a broad range of social problems, and he saw agriculture as an important and fundamental dimension of social culture.

As with Christoph, Gary has reservations about the typical CSA concept. "The problem with community supported agriculture," he says, "is that you have high concentrations of people in the cities and then these CSAs out in the country. Well, where are the people in the cities going to get their food? How can they work in a constructive way with agriculture? There has to be an association of food consumers in some way.

"The CSA is a method by which the farmer is assured a market, and that takes care of one of his principal worries: somebody to sell to, and a fair price for his products. Also the farmer has money in hand to invest in the land, and that, of course, is another worry. It's certainly better from the farmer's perspective than taking a loan at the bank and then having to pay interest on the money for this year's crop. But the real question is how do you expand this? It works on a local level for the local farm, but we have to find a way to make it applicable to agriculture in general.

"Here in the Northeast the winters are long and the growing season is short. People can support a farm in a region, but what about the farmer who grows oranges for you in Florida? Or lettuce in February in Arizona? How can you support these other farms, too? People are going to eat bananas and oranges and other things out of season. Community supported agriculture does address many key issues on a local level, but the percentage of agriculture that it can deal with, at least as far as the form is presently conceived, is limited. I think there's ways that can be found to get beyond this. And we must. We can't be selfish and just support our local farms; we have to broaden our response to other farmers . . . In terms of the farms of the future, whatever we develop has to embrace all of agriculture. We need a broader context than at present or offered by the CSA concept as presently formulated."

Based on his experience, Gary believes that in the future there has to be a place for the checkout counter within the CSA concept. He sees the checkout counter and the choice that it implies as good ideas. Some people involved with CSAs think stores and their checkout counters are dispensable elements in getting food from farmers to consumers. But Gary believes they may not be. At Hawthorne Valley, the retail operation not only supports this farm, but also dozens of other farms and producers. "When you have a CSA," Gary says, "it supports one specific farm. There is a place for a store that supports a number of farms and a wider variety of goods."

Gary has some pointed advise for people who are attracted to farming as a way of life: "Get away from wholesale and focus on retail sales. That puts you in direct contact with consumers, which is critical. That's a beginning, to create some kind of understanding, to give the consumer and the producer a chance to enter into a dialogue. From there the relationship might evolve, the consumer might have some land, or some money they want to invest; or they might come to support you more directly, as with the CSA movement."

Get the Government Out

When it comes to the role of government in agriculture, Gary Lamb has an unyielding point of view. No doubt his position crystallized in 1988 and 1989 when New York State began thinking about banning raw milk. Hawthorne Valley Farm is the largest producer of raw milk in a state where there are only about seven producers altogether.

When the move toward banning raw milk intensified, the farm saw it as an opportunity to look again at the relationship of the government to the farm. Gary says that, ultimately, all agriculture is community supported unless the government is involved. "Many agricultural problems are due to government intervention, especially in the 1980's and now in the Midwest, where you have all of these farms going out of business. They have been taking out huge loans, and have been directed by the government to use chemicals in various ways, and subsidized to grow things that people don't really want or need. The enthusiasm, or connection to agriculture has been diminished. If you are growing a crop that you know people aren't going to eat, that is just going into storage somewhere or that will be dumped out so that you can get paid a subsidy, then there's no real connection to the earth. No human being will develop a connection to the land or the soil and go to all the effort it takes to grow food and then just watch it be thrown away. Government intervention introduces elements that have nothing to do with agriculture.

"Government ought not to be involved in agriculture at all. If there's going to be a Department of Agriculture, then it ought to be run by farmers. After all, what does a politician really know about agriculture? Safety and other issues may be appropriate for government regulation, but when you are talking about technical expertise with crops, or the relationship of the farmer to the consumer, then history clearly points out that the government has botched it up. This is a crucial issue. Historically, if you look at the worst situations that farmers have had to face, you'll see that they are somehow connected with government involvement and support.

"Let's assume the farm of the future is going to be a low-input, non-chemical organic or biodynamic farm. What has already happened in the organic and biodynamic circles is that they've supported, or are even pushing to get government standards for organic farming. That's the last thing to do. There's enough history to prove that getting mixed up with the government leads to no good.

"I don't understand why people who are involved with alternative agriculture would go to the government and ask them to legislate the standard of organics, because then it will become a bureaucratized definition and it'll probably be enforced by someone who has no connection to agriculture or organics. I see problems there. The whole certification of whether something is organic or biodynamic I think should be dealt with by the private sector, by organizations with their own standards. Besides, one organic standard through all the regions of the United States doesn't make sense. What you'll have is an inflexible set of rules in the name of consumer safety and legitimacy. But that's just a way of taking the life out of it."

EXAMPLE 6
Tax-Supported Farms

One of Stan White's abiding pleasures is to show off his Border Collie, Hans. Directed by a hand signal, Hans will streak across the rolling pastures of Codman Farm with obvious delight to herd sheep and goats. The pastures where Hans earns his keep are in Lincoln, Massachusetts, a wealthy suburb of Boston. Since land prices in Lincoln can verge on an astronomical $100,000 an acre, those pastures might well have been bulldozed for a shopping center, or condominiums, or an office building. But because the townspeople wanted to preserve the tradition of farming in their community, they took the steps necessary to save the land.

Codman Farm is one of several tax-supported farms which took root in the suburbs of Greater Boston during the 1970's and 80's, some to preserve open space in crowded suburbs, and some to create educational and recreational opportunities for young people.

Three other projects near the Codman Farm are similar: Land's Sake in Weston, which produces flowers and vegetables on 35 acres of town land and contracts with the town and with private landowners to manage open land; Greenpower Farms, also in Weston; and the Natick Community Farm. All are tax-subsidized agricultural projects designed to involve local youth in the land and, in the case of Greenpower, to provide fresh low-cost food for the residents of the inner city.

Each of these farms employs ingenuity to preserve a measure of agricultural heritage in areas where farming has been all but eradicated.

The Problem of Land

Twenty-five years ago Lincoln was an active agricultural community. There were many working dairy farms, but as the urban sprawl spilled out of Boston, farmers were priced out. They either moved west or north, or just closed up shop. At the same time this was happening, some people in town had the foresight to realize that sooner or later all the farmland was going to be converted to house lots. And so they started purchasing as much farmland as they could get their hands on for conservation purposes.

Some of land they bought was good for agriculture, but a lot of it was marginal: good for hay, or pasture, but not Class 1 land where you could plant row crops. Since all the dairy and livestock farmers had moved out, Lincoln found itself in a predicament. It had a lot of farmland which was not being used because there were no farmers around. That was problem number one.

Problem number two arose when a woman named Dorothy Codman passed away and left a small farm to the town in 1969. All of a sudden Lincoln found itself owning a barn and 16 acres. So it started doing what most towns would do in similar circumstances: using the barn to store junk tires, police car doors, pieces of wire, and highway equipment.

Many people in town recognized that this was not the best use for the property, so in 1973 they voted to turn the place back into a farm. A group formed a non-profit, educational corporation and called itself Codman Farms, Inc. The town, via town meeting, granted them the right to lease the land and barns for use as a farmstead.

As it was set up, the farm is financially independent of the town. The town does provide money for major, big-ticket items because they are the owners and landlords. For instance, if Codman Farms needs a new roof on the barn they go to Town Meeting and ask to have the money appropriated. But if it's small maintenance, like a broken door or window, then they do it and pay for it themselves. Likewise, the farm corporation pays for all routine operating costs, including utilities.

After the farm board established itself, it sought out a farmer, eventually hiring Stan White. A graduate of the University of Wisconsin with a degree in agriculture, White has an interest in combining traditional and modern farming methods — a blend that has worked well at Codman Farm.

The 12-member board meets monthly to oversee the farm. They have ultimate control over the pursestrings. But like any executive who runs a business, Stan White, as the farmer, makes the day-to-day decisions.

There are three chief motivations for the members of the Board to volunteer their time and expertise. If the farm were not operating, over 130 acres of land would be unmanaged and possibly for sale on the real estate market. Codman Farm is also the only viable farm in town. Some other people farm part time, but Stan White is the only Lincoln resident who earns all his income from farming. So if the farm should go, Lincoln would, for all intents and purposes, lose its agricultural heritage. Finally, the open land is scenic; by its very beauty it helps to keep local property values high.

A Patchwork Operation

As presently constituted, Codman Farm is a patchwork quilt of about 98 acres. The nucleus is the farmhouse where Stan White lives, the barn, and 16 acres of pasture. But there's a lot of other land under his care, most of it scattered around town, some fields as much as three miles away. In addition to farming the 98 acres of town-owned land, Stan hires himself and his equipment out to do rototilling, plowing, and hay cutting around town. The heart of the farm is the livestock operation, which is highly specialized. Stan has become perhaps the top expert in New England on rare and endangered breeds of cows and sheep.

In general, Codman Farm runs at a break-even rate. The annual budget is about $90,000 to cover feeds, fertilizers, labor, livestock, equipment, and so forth. About 90% of the income derives from the farm's operating receipts, and about 10% comes from donations. At the end of the year the ledger balances, but as Stan White comments, because the farm does not make enough to reinvest manures in the land, in the long run it may turn out to be a losing operation.

Stan gets a salary and subsidized housing in the farmhouse. But he has no benefits, and no equity in the property. So why does he do it? He says he loves the work and also he has a deep commitment to raising endangered breeds of livestock. Basically, his ideals have kept him going.

The biggest cost of the farm operation is not the manure or fertilizer or even the livestock, but the labor. "I pay $6.50 an hour for people to throw bales of hay," he explains. "The workers figure they are getting ripped off because they could go flip burgers instead for $8 and get a health program, too. They don't get that here. You gotta do it because you enjoy it. That's the biggest crunch we face."

Community Support

Without community involvement and non-profit status, the farm could not keep running. The land values in Lincoln are just too high. But by the same token, Codman Farm is fortunate to be in an area where it can get away with charging premium prices for their products. "There's no way a person's going to get $7 for a pound of veal in the Northeast Kingdom of Vermont," Stan White says. "That's a given. But then again, in the Northeast Kingdom you don't have to pay $6.50 an hour for someone to pick up your hay."

Community support comes from the Board of Directors in the form of advice on accounting and taxes, adding much sophisticated advice to fine tune the financial side of the operation. Stan White is particularly grateful for this support: "it's the kind of advice every farmer should get, whether it comes from a board or a consultant. Most farmers are not innately salespeople and most farmers are not naturally astute accountants, you know, thinking of tax laws and depreciation, and so on, every time they have to make a decision. So it's nice to have support in that. I can do all of that stuff pretty well, but when the growing season starts, and push comes to shove, then I find that you end up with the tasks which really have to be done and which interest you the most. And as you become busy, the business consciousness can slip away because the farm work has to take priority — it has to be done. I find I can't make a really good business decision in July or August because I'm just too physically and mentally tired to weigh the facts and step back and ask 'where are we going?' The only time I can do that is in the winter."

Central Questions

As Stan White sees it, one of the central questions of the 1990s is whether we are willing to pay the farmers the cost of growing clean food. "I don't like being subsidized. To me, there just seems to be something intrinsically wrong with it. But the facts are that, to one degree or another, farms have to be supported, not subsidized, with a financial input. If farmers got a true price for the cost of production, they would not have to be subsidized. But for various

99

reasons that does not happen. I'd like to see the farm carry its own financial weight," he says, "but everything is so twisted out of proportion that you need extra money coming in."

Undoubtedly, Codman Farm is an unusual operation with a generous measure of community support. The town voted to use the land as a farm, and a board of citizens came forward to create and help manage it. But is it workable elsewhere? Stan White says yes, it can be workable if you strike the right balances. "First, you have to realize that a person needs to get more value, or cash from the products of the farm just to have economic survival. He's got to break even and get a reasonable pay for his labor, because no matter how noble he is, if he ends up two notches above welfare level, then what's the point?

"The worst thing is that, with an arrangement like this, since the farmer does not own the property he has very little incentive to hang around other than to collect his paycheck. If you live, for example, on your own Vermont farm and you end up working your rear off, literally 3,000 to 3,500 hours a year, just to make $500 or $1,000 after expenses, and sometimes to even lose money, well then at least you still have equity in the land. But that doesn't happen in a situation like this, and that has to be put into the equation, too. I have a strong feeling of land stewardship. It's important to me. But that just happens to be a dimension of my character. It's not because this land came down to me from my grandfather and will go on to my grandchildren. That's not going to happen. You may not always get stewardship of the land. You have to have unusual people and unusual balances to make it work, realizing that this kind of farming is just plain going to wind up costing more because you are going to have those management problems which you don't have if the person owns the farm straight out. I think the most economical and most successful way to go would be to have the farmer keep on owning the land, or at least to have some equity involvement. You would see the difference."

The Natick Community Farm

Located in a densely developed suburb of Boston, in a town with only 7% of the land left in open space, the Natick Community Farm has also become a model of tax-supported agriculture.

The farm began in 1976 when the town used federal CETA (Comprehensive Employment and Training Act) funds to start a garden project that was primarily intended to keep school kids off the street in the summer. After the first year, the program director found the 22 acres where the farm is now located and, with the help of the school kids, convinced the local school committee to purchase the land.

The school committee acted with the thought that the land was a good investment in case they needed a new school sometime in the future. But they have continued to grant a series of five-year leases to the non-profit, education corporation that has been formed to run the farm. Then, toward the end of the 1980's, the farm's board of directors began the process of registering the land under the Massachusetts Agricultural Preservation Act so that it will remain in agricultural use on into the future.

Farm Director Linda Simpkins says the Natick Community Farm could be a model for many other communities interested in maintaining agriculture and open land. "Anyone in town can come out and use this land as long as they are respectful of it," she says. "So far the community really seems to appreciate that."

The farm is run like a homestead operation, with a small number of sheep, cows, goats, chickens, pigs, turkeys, ducks and rabbits. They have five acres of organic vegetables and fruit. And they run educational programs throughout the year for all the children in town, grades 1 through 5. The income from the sale of vegetables, fruit, meat, maple syrup, milk and eggs comes to about $16,000 a year. That money is used to provide low-paying, educational summer jobs for 130 children each year.

The farm gets $38,000 a year from the town's Department of Recreation and Human Services, and $31,000 in donations is raised by the farms Board of Directors.

Associate Director Martin Gorsky says "what we feel we are about for this community is a link to the natural world. Over 14,000 people come through here each year — students, consumers, educators, nature lovers, and so forth. In a crowded town, many, many people feel the need for this place and the open space it provides. Lots of people have lost their link with the land, and the Natick

Community Farm helps to reestablish that link through the programs with the children, and just by being here."

Linda Simpkins and Martin Gorsky have started to meet with the workers at nearby community farms in Lincoln and Weston to discuss common problems, and for mutual support.

Greenpower Farm

Back in 1970, in the prosperous Boston suburb of Weston, Bill McIllwaine spearheaded the effort to create Greenpower Farm, an unusual operation that grows produce and sells it at reasonable prices to low-income people in the inner city. The idea for the farm arose out of the recognition that low-income people often do not have access to farm-fresh produce, and the idea caught on.

"We've expanded every year," Bill explains. "By the end of the 1970's we had over 20 acres in cultivation, and dozens of people involved." In fact, after the farm's first year of operation, every resident of Weston became involved in a way: slowly at first, then more steadily, the town began to support the farm with funding through its Youth Commission. Since 1973, Greenpower Farm has received about $65,000 annually from the town. That amount is supplemented by sales of vegetables and maple syrup.

"We pay a small wage to the kids in the summer, and they get a tremendous amount out of the program," Bill explains. "We've never made a profit, but we've been able to keep on going . . . Lots of people criticize the program and say that only a rich town could do this, but I disagree. When you think about it, $65,000 is not much for a good sized town, and there are tremendous benefits as a result."

Weston has one other community supported agricultural enterprise, Land's Sake Farm. It relies on CSA-like memberships to meet about 10 percent of its $100,000 yearly budget. Most of the rest comes from the sale of firewoood, maple syrup, wool, flowers, and vegetables.

EXAMPLE 7
A Community Market:
The CSA Concept Meets the Grange

Just down the road from the Temple-Wilton Community Farm in New Hampshire, a small group in the town of New Ipswich has begun an experiment that adds another dimension to the CSA concept. As a step toward eventually creating a farm, they are forming an organic farmer's market in cooperation with the local Grange.

The New Ipswich CSA group began meeting on June 14, 1989. It took them three meetings over three months to recognize that the hope of having a CSA farm up and producing a year later, in the summer of 1990, was just too ambitious. They had no suitable land, and though most members of the group were seasoned gardeners, no one had experience farming. Realizing that it is a major economic and technical jump from a backyard garden to a farm that can reliably feed a large number of families, they began considering other possibilities.

As they met over the summer of 1989, the New Ipswich CSA group evaluated its resources: a membership of skilled gardeners, general commitment to the CSA concept, and the support of the local Grange. The 115-year old Watatic Grange, a chapter of the National Grange organization, had several valuable resources that it was willing to make available to the CSA: the stability of a mature membership, a beautiful old building located on Main Street in the village of New Ipswich, and the goodwill it had built up in the community over many decades of supporting local agriculture, homemaking, and crafts. While the Grange Hall needed repair and refurbishing, it had the advantage of a prime location. Eventually, the CSA group decided, as its first step toward eventually establishing a farm, to create a farmer's market where local gardeners could sell organic produce, eggs, and crafts to residents of the town.

To clearly show its linkage with the Watatic Grange, the CSA group decided to name itself the Watatic Community Farm. The goal was to eventually create a full biodynamic farm, but to start with

103

a farmer's market serving the towns of New Ipswich, Greenville, and Mason, New Hampshire.

As of this writing, the market idea had not yet been fully developed, nor had it been put to the test. Since it was yet another alternative support of the general CSA concept, though, the authors decided to present an outline of the plan to stimulate the thinking of others.

Over the winter of 1989-90 the Watatic Community Farm began to line up gardeners and others who produced agricultural goods, including those who want to raise chickens, pigs, and beef cattle. The group defined its goal of establishing a roadside farmer's market in front of the Grange Hall on Saturday mornings in July, August, September, and possibly October. The market would be a place where local growers could sell their cucumbers, tomatoes, raspberries, and so forth. And it would also be a meeting place where people who were interested in buying a share of a pig, or a quarter of a beef cattle could meet and transact business.

In a time of tightening economics, with an increasing number of layoffs in the area, the Watatic Community Farm members felt that the market could eventually also provide an important venue for trade and barter among local residents.

To develop rules and regulations for how the market would work, the Community Farm members turned to the Massachusetts Department of Agriculture (100 Cambridge St., Boston, MA 02202), which has published a booklet entitled *How to Organize and Run a Successful Farmer's Market* by Julia Freedgood. The book, along with sample copies of rules and regulations from other markets, helped the New Ipswich group make the necessary preparations for its first season of operation (see Appendix D).

As with all such undertakings, there were many questions to resolve. How much of a fee should the market charge growers who wish to sell? What permits are necessary? Would the market accept food stamps? Should the CSA incorporate? How could it promote its activities? Who would supervise the market to be sure it's orderly?

As the concept developed, the Watatic Community Farm decided to give a percentage of its profits to the Grange in recompense for the use of Union Hall, and also to support the ongoing work of the

Grange in the Community. Several members of the Community Farm also began to consider applying for membership in the Grange.

Nuts and Bolts

The Watatic Grange was established on September 11, 1874, and named after a nearby mountain with a panoramic view. At the time the Community Farm group began meeting, the Grange had 22 members, but only a fraction of that number remained active. The Grange generously extended the use of its meeting hall to the fledgling CSA group, and over time mutual interest began to develop.

As the members of the Watatic Community Farm began to address the nuts and bolts issues of establishing a Farmer's Market, they decided to form two groups: a growers group, and a consumers group. The growers group would undertake the critical task of lining up and coordinating the local gardeners and farmers who would grow the produce for the market. This involves recruiting the growers through local notices in the paper, direct contact, and the grapevine. The growers would specify what crops they plan to grow, the amount they intend to produce, and an estimate of when those crops would be ready for the market.

Without a steady supply of a broad variety of crops, the market would not be able to build a strong and reliable base of consumers. The feeling was that, from the outset, the Farmer's Market needed to have a solid presence: it had to be there each Saturday morning, and to offer a sufficient variety, quantity, and quality of foods to attract and hold consumers. The growers group would also set prices, and work out the details of the bookkeeping. The initial plan was to allow free market pricing, but to encourage growers to set fair prices using as a guide the *Weekly Market Bulletin* issued by the New Hampshire Department of Agriculture.

The consumers group would take on the responsibility of developing the organization: meeting with town officials, arranging newspaper stories and notices, publishing a quarterly newsletter to keep people informed, and holding the network of consumers and growers together with clear and regular communication.

A Community Focal Point

As we enter the 1990's, the CSA market concept offers a helpful alternative. Members of a community may be unable or unwilling to farm, but with minimal training they can begin to garden, and to garden extensively so that there is a surplus which can be shared with the community through a market. Depending on how it is set up, the market can ensure that gardeners and farmers get a fair return for the produce they are sharing with the community. And if cash money is scarce, the markets can serve as a focal point for barter.

Since local chapters of the Grange already exist in many American communities, it is uniquely suited to the challenge of making such markets work, and providing a source of fresh, clean, locally grown food.

The Farmer's Market concept also offers a gradual step for a community group that wishes to build a network of relationships that might eventually support a farm with a land base.

GUIDELINES
The Basic CSA Concept:
Some Guidelines for Getting Started

In its starkest terms, Community Supported Agriculture (CSA) is a concept describing a community-based organization of producers and consumers. The consumers agree to provide direct, upfront support for the local growers who will produce their food. The growers agree to do their best to provide a sufficient quantity and quality of food to meet the needs and expectations of the consumers. Within this general arrangement there is room for much variation, depending on the resources and desires of the participants.

If there is a common understanding among people who have been involved with CSAs, it is that there is no formula. Each group that gets started has to assess its own goals, skills and resources, and then proceed from that point. The decisions any group reaches, and the challenges it faces will vary from case to case. Still, though, there are some general guidelines that should be considered by any group starting out.

• Begin by sharing the idea informally with a small number of people. When several people are interested, call a public meeting. Announce it through posters in local businesses, notices in the local newspaper, and elsehwere. You may even want to consider the advisability of inviting officials from the local government, including members of conservation commissions, planning boards, and governing councils. Steadily build public awareness of and support for what you are doing. The first public meeting should present a clear overview of the possibilities, allow time for general discussion, and include a sign-up sheet to collect the names and addresses of everyone who is interested.

• Form a core group. For a CSA to get off the ground, it must have a committed core group of five or more people who will regularly attend meetings, and do the work: make copies, place phone calls, execute decisions, and so forth. When idealistic movements like a CSA get started, they often attract many curious people. But of the initial crowds, in general only a small number of people will actually continue to come to the meetings and do the

107

work necessary to organize and operate a CSA. Without a minimum of at least five people, the core group will quickly exhaust itself and come to feel discouraged.

• Build a list of names and phone numbers, and pass a hat for donations. The core group may well have to donate some stamps and phone calls at the outset, which is something most people are willing to do. But if there is no provision for incoming resources from the wider community, the core group may soon feel depleted.

• Develop a clearly defined vision, with specific goals and a practical business plan for meeting the vision. It may take many meetings to define the vision and the plan, and they are likely to change over time, but these are essential and healthy parts of the process. If someone attempts to establish a CSA as if it were a rigid franchise concept, then other people will have been denied a chance to contribute to and build the idea. Consequently, they will have no sense of belonging, and perhaps a low level of commitment.

• Foster a democratic process based on consensus. A CSA may want to rotate the chairmanship from meeting to meeting in the beginning. Having a different facilitator for different meetings ensures that the power or authority to make decisions does not become vested in the hands of one or two members, but is instead widely shared. Such democratic processes ultimately strengthen the sense of membership in the group, and help to prevent individuals from overextending themselves.

• Have a well-defined agenda for each meeting and stick to it. Whoever is going to chair one of the organizational meetings should, at the outset of the meeting, define what the agenda is and the order in which items will be discussed. These straightforward business procedures help ensure that the creative energy of the group members is clearly focused.

• Share responsibility. The more members who are carrying the weight of the CSA and participating in decisions and operating activities, the greater the chances of success.

• Be systematic. The CSA concept arises out of the realm of ideals, but it must be based in practical and efficient business sytems. Computers, for instance, can be a tremendous help in keeping track of members, finances, and crop estimates. While farmers and gardeners may lack such skills, other members of the group may be able to offer assistance with computers, business practices, and other technical skills.

• Make the basic financial decision about how member families will participate. Some CSAs divide the entire cost of running the farm, then use this as a basis for assessing the cost of a share. Other CSAs, such as the Temple-Wilton Farm, depend on membership pledges, like public radio or TV stations. They work out their budget over the winter and present it at a shareholders meeting. Shareholders then make pledges. If the total of the pledges falls short of the budget, the shareholders are asked to increase their free will pledges until the anticipated expenses are met.

• Establish a land base, either with a long-term lease or through a land trust. While the acquisition of land may take a number of years, it is essential for the ultimate success of a CSA. If a CSA is forced to rent or lease land for growing from year to year, then those who work the land will have less motivation for making long-term improvements and investments in upgrading the quality of the soil and the physical infrastructure. Additionally, the general membership of the CSA will have no focal point on the land for its energies, something that is essential if the CSA is to maintain a sense of stewardship with the earth.

• Publish a newsletter. Virtually every existing CSA issues a newsletter, generally on a quarterly basis. The newsletter serves as an essential communications link to keep members informed of the decisions and developments that concern them. Without regular news of the CSA, members may lose touch and then lose interest.

• Contact your state Department of Agriculture and let them know what you are doing, while finding out what resources they make available to farmers and farm groups. Many states offer resources or services that can be of help.

• Be patient. It takes time to make any community undertaking work. If a CSA is going to be successful, it will have to evolve gradually over time. But if it's worth doing, it's worth taking the time to develop it.

PART III

Appendices

APPENDIX A
History of the Temple-Wilton Community Farm
1986 - 1990 in Documents (including a sample budget)
Trauger Groh

Following are documents and excerpts from the monthly Newsletter of the Temple/Wilton Community Farm, that show some of the history, the motivation and the problems of this endeavor. Although the form of cooperation on this farm is specific to this place, questions have arisen here that will come up in other community-supported farms as well. Using the material as it has come up in the dialogue of the people involved may give this description a certain freshness and un-academic style.

The lifebreath of any cooperation of people on and around a farm is the dialogue while working together, in regular or irregular meetings, and through a newsletter that covers the actual situation of the farm and its problems. It should give food for thought about the principles of this cooperation, about the why and the how of a life-filled farm organism and the food, the learning, and the social involvement that is connected with it. A newsletter is of great importance for any such operation, because not everyone who feels connected to the farm is willing and able to attend meetings regularly. Beyond that, its content can be used to introduce new members and to help locate new Farm Communities.

When people share the costs of a farm in the way it is done on our farm, they give, on the one hand, the group of active farmers

110

the freedom to act out of their innermost understanding of the natural processes of the farm — not looking for a short term monetary profit, but just aiming at quality on all levels. On the other hand, they can expect that the acting farmers give reason for what they are doing and for the outcome of their doings whether they are good or bad. The "deal" can be described as follows: "We want to relieve you — farmer — from instant financial pressure if you promise to go only for the highest quality in the future and if you introduce us into the secrets of this farm and the reasons in your work." This reason begins with a thorough projection of the future work and its expenses.

Every farmer should be able to work out a detailed and true budget. This is not only necessary for a supportive community, it is first of all necessary for the farmer himself, for his clarity about his intentions. A diversified farm with various livestock and many crops needs a real overview and a thorough planning of everything. Here it is not, as in conventional agriculture, the cost-effect analysis that is fruitful. It is the insight and explanation of how any measure affects the fertility of the total "farm organism." A cost-profit analysis for keeping a few beehives on the farm or a small flock of sheep is worthless and usually impossible, but we should train ourselves to understand and to explain the importance of these parts of the farm for the benefit and health of the whole farm.

Following an agreement on such a budget we have then to demonstrate and explain the successes and failures of the year on the farm. First, the farmer, and through him everyone who wants to carry responsibility for the farm, has to understand why certain things went this or that way in the crops and with the animals this year. The success — or failure — of one year has now to be measured by our intentions. If our intentions are not to go for profit but for an "ever growing diversity on the farm adapted to the needs of the local population, while the import of substances and energies from outside the farm into the farm permanently decreases," if we have this as a goal, then we have to be able to document the progress on this road at any time and every year. The whole farm community with all its members (75 families in the case of the Temple/Wilton Community Farm) should ideally go through the following process together:

1. Coming to an understanding of the idea "farm organism" for this special location with its deficiencies and properties in soil, climate, farm structure.

2. Receiving from and discussing with the active farmers their plan for the year to come closer to this idea, together with the expenses that are necessary to achieve it.

3. Helping physically and by supporting the expenses to implement that plan.

4. Learning the details that occur in the year for a better understanding of this farm.

5. Looking back on the achievements and failures and measuring them against the idea.

All these steps can be found in the excerpts and documents that have been laid down and are included here. We find articles that deal with the basic ideas spiritually and economically. We find something about the development of the farm from year to year. We find reflections about the money flow in the farm and financing questions. We find the question: "How can we get the work done properly without hired labor?" And we find out details of the difficulties with certain crops.

To have a realistic approach to the question of budgeting we will include a typical annual budget of our farm.

First, a short review of the situation in the area of the towns of Temple and Wilton, New Hampshire. The region west of Boston, Massachusetts, and especially west of Nashua, New Hampshire, was in 1986 in rapid development. High tech industries in Massachusetts and southern New Hampshire drew more and more people into the region. They needed house lots and shopping malls and caused the prices of land to skyrocket. Due to the minor quality of the farmland around here, many farms had been given up long ago. The land had bushed or wooded in. The new intensive development pushed out many of the few farms that had remained. In and around Wilton, people had settled who were looking for free and independent schools for their children: the High Mowing and Pine Hill Schools. Amongst those people were some who were deeply concerned about the "life" value of the supermarket foods and of the decline of farming in this formerly rural area.

In the years before 1986 two farmers had tried to establish organic farms in the area to meet the needs and wishes of this local population. The capital cost and the immense cost and labor that comes out of clearing and recultivating abandoned, bushed-in farmland made these efforts not viable, but out of these efforts a small core group had evolved farmers and non-farmers alike, who seriously wanted a true farm in the region and were discussing ways to achieve it.

With a new little farm being acquired in 1985 and a new farmer having arrived, this group could now realistically discuss the social and economic implications of establishing and carrying a farm organism. Two small farms with together not more than 20 acres of open land and the possibility of clearing some more land were the beginning. Part of the land of one farm belonged to a foundation that mainly was engaged in caring for and "lifesharing" with adult handicapped people.

The frequent meeting of about 20 families in the springtime of 1986 brought about the will to start a farm together on the first of July 1986 and brought about a document of intent that has remained so far the only basic document of this initiative trying to describe intentions and procedures of the new farm. It reads like this:

> The community farm was born out of the desire of a group of farmers and gardeners to unite their efforts and their land into one organism, in order to serve the local community with bio-dynamically grown food. The first meeting took place in February and March 1986, and out of these came forth some basic aims and concepts:
>
> Landholders — The landholders give the members of the community farm, individually and in association, the right of agricultural use of all their land, farm buildings, farm animals and farm machinery, except their homes and house gardens. This right is given free of charge, but with compensation made for all costs of the property: land taxes, insurance, depreciation and repairs.
>
> Farmers — By taking over the right of use of the above-mentioned land, and with it the responsibility for the good agricultural use of this land, all members of the community become farmers. They either enact their rights to farm directly, by actually planning and doing the farm work, or by letting those members who have the time and skills do it, farm in their name. Those members who do the

planning and the farm work on an ongoing basis and as a main occupation are called the "Active Farmers."

Aims of the Farmers:

1. *Spiritual Aims* — To make life on Earth possible ever again and every year anew in such a way that both the individual and mankind at large can live towards their spiritual destiny.

To make land-use and working of the land a way of self-education, and education in the sense that a better understanding of nature can lead to a better understanding of man.

To create the farm organism in a way that the above becomes possible and that it is made available in a therapeutic way to those who suffer from damages created by civilization and from other handicaps that need special care.

2. *Legal Aims* — To make access to farm land available for as many people as possible.

To create forms of cooperation that exclude employment and any form of paid labor.

3. *Economic Aims* — To develop a natural organism so that it reproduces itself better and better and becomes more and more diversified, so that it can be a primary source of food for the local community.

To achieve that reproduction and diversification with the help of the forces of nature inside the organism so that it becomes less and less necessary to introduce into the organism substances and energy from outside and so that human labor is used as economically as possible.

Individual profit through farming is not an economic aim of the farmers.

The farmers agree on certain principles to make the cooperation in the agricultural community possible:

1. Everyone is individually fully responsible for his or her doings and its consequences. To enable others to help him in his initiatives and the consequences thereof, the individual must, in a timely way, let the others know what he intends to do.

2. The individual incurs expenses to serve his initiatives. The expenditures made by the single individual increase the costs for all the others. Therefore, the individual, in cooperation with other

114

individuals, has to declare what costs he projects to fulfill his initiatives within a certain time. The declarations of those that intend to spend money, combined together, make the annual budget. This budget has to be approved by the assembly of farmers (all members). Once the budget is approved, the single farmer is free to spend, throughout the year, the amount of money he has in the approved budget.

Every farmer who spends money is ready to keep books and records of such expenditures. The farmers agree on a scheme of categories in which the expenses are accounted for. The books have to verify annually how far the economic aims have been achieved.

3. Every farmer (member) can leave the Community at the end of the year when he has paid his part of the annual cost.

4. Every farmer gives all the other farmers of the group the right to substitute for him in his work if he fails to do or complete something he has taken on.

5. It is understood that the better the cooperation between farmers is working, the less goods and services are brought into the farm organism by individuals at the expense of all others. It is understood that the least desirable thing that should be purchased from outside is human labor.

6. All farmers have to take care that they spend enough time on observation, planning, and communication.

7. The motivation to do things on the farm should come more and more out of the spiritual realm and less and less be directed by solely financial constraints.

The above statement of aims and concepts became the focus of a series of intense meetings between potential "active farmers" and other potential members.

These meetings are reported in the following:

Meeting on Saturday, April 19th: Lincoln Geiger presented the idea of a united farm effort between Echo Farm, Crossroad Farm, Temple Gardens and Temple Road Farm, as it had been worked out at the weekly meetings of those who work on these places and others who feel themselves connected to this farm impulse in Wilton/Temple. Lincoln explained that any of these places taken by themselves are too small to create a real diversified and healthy farm organism that can be managed according to the advice Rudolf Steiner gave for agriculture at Whitsun 1924. In size Crossroad and Echo Farm together represent 65 acres, Temple Gardens 7 acres, Temple Road Farm 33 acres. Of these 105 acres total, 20 are cleared and in agricultural use this season. 24 more acres will be cleared this year to come into use in the next 2-5 years, 6 acres is a rented apple and blueberry orchard, 55 acres is woodland, ponds, buildings and home gardens. Therefore, including the rented orchard, we now have 26 acres in production and expect a total of 50 acres within the next few years. This acreage is now, and will perhaps even more so in the future, be supplemented by land that is available to the group through short-term lease arrangements.

The new farm organism will be established around a dairy herd that eventually could consist of 12 cows, 1 bull and the number of heifers and calves that are necessary to keep up such a herd (approximately 10 heifers and calves). At present 4 milking cows and 4 calves constitute the herd. This year at Temple Gardens and Echo Farm there will be together 1 acre of vegetables. Another 1.8 acres of vegetables will be growing at Temple Road Farm. Six acres are in fruit. Eleven acres are grazing, hayfield and fodder crops. A small flock of sheep is kept now at Crossroad Farm. In the future more sheep will be kept and beginning this year, poultry. The little land that is available this year does not permit us to grow grain. Another obstacle in growing baking and fodder grains is the lack of grain harvesting equipment in this part of the country. This leads us also to a lack of bedding straw and to problems in the handling of the cow manure. We hope therefore to grow some type of grain in the future and are presently studying this problem to find good solutions to both variety and machine questions.

Lincoln also spoke about the future possibility of processing milk, vegetables and fruit.

Saturday, May 3rd: In this meeting Anthony Graham first named those persons who will take on the long-term responsibility for planning and working on the new farm, in their own names and in the names of those others who choose to become members of the Farm Group (these being able to farm only occasionally or not at all). These Active Farmers are Lincoln Geiger, Celine Gendron,

Alice Bennett-Groh, Trauger Groh and Anthony Graham. These Active Farmers have worked out for this first year how much money they intend to spend in their farming activities and these figures have been brought together into a budget. This budget was distributed and discussed in this meeting. It shows planned expenses of approximately $23,000. This includes operational costs only since the farm community has chosen not to deal with any capital investments. The budget was characterized as an "emergency budget" since it only allows the most basic and necessary expenditures to be made in this first year. The budget does not include any farm building repairs, or depreciation payments on machinery or buildings. With the kind permission of the respective proprietors, these payments must be delayed at least until next year's budget. The current budget does include the partial private or household income for one of the active farmers.

On the income side, the estimated income for the sale of milk, vegetables and fruit, based on the experience of last year, was set at approximately $18,000. This left a deficit of $5,000. How the members of the Farm Group would choose to deal with this deficit was left as the topic for the next meeting. As a matter of equality and accurate record keeping, it was decided that every member of the Farm Group, including all the active farmers, would pay for everything taken from the farm production.

On Saturday, May 10th a new approach to carrying the costs of the farm and enjoying its goods and services was brought forward and largely approved. It was suggested that every member household pay a certain sum per member of the household on a monthly or annual basis to cover the cost of the annual budget and that then the produce of the farm, the vegetables, fruit and milk could be taken by the members free of charge. In this way, carrying the operational costs of the farm which are created out of the desires and initiatives of human beings, and the enjoyment of the produce that comes as a free gift of nature are seen and treated separately.

The participants of this meeting and all interested people were and are asked to come forward with an estimate of what they could contribute per month or per year to cover farm costs. It was mentioned by one person that if 20 households with 4 members each, or a total of 80 people participate at a rate of $24 per month ($288 *per year), the farm budget could be covered. But since the income of households vary so widely, it was felt that the contributions would and should also vary. Lower contributions will be possible due to the compensatory effect of farm sales to non-members at higher market prices.*

The discussion has begun and will continue regarding the method of distribution of the farm products among the members, and of the feasibility of including the products of other Biodynamic farms in this process. A number of detail arrangements in bookkeeping and distribution await discussion in the May 24th meeting. Please come with your ideas and a good sense of what you are able and willing to contribute.

Eventually a formula was worked out whereby all raw farm produce would be available to members free of charge if they could cover the proposed budget through monthly contributions over the course of the year. This idea seemed to be a breakthrough in severing the direct link between food and money and was accepted with enthusiasm. Financial pledges were then made, based on the ability to pay rather than on the amount of food to be taken, and having made a contribution, the member was free to take as much food as was needed, dependent, of course, on availability.

The first budget was about $23,000. It was squeezed as far as possible and somehow built on personal sacrifices of the Active Farmers, who wanted to make possible the beginning of the new farm under the most difficult circumstances; on little land (20 acres) and this land acid and undeveloped, low in humus.

And now it started:

We are pleased to announce open hours for the Community Farm Shop. FOR VEGETABLES: The shop will be open and there will be someone there to help you on WEDNESDAYS 2-5 pm and SATURDAYS 10 am-1 pm. Our first shop day will be on WEDNESDAY, JULY 16. For the next two weeks we will have lettuce and some bunching onions at the shop, and pick-your-own peas at Temple Gardens. FOR MILK USERS: Lincoln is now working out the schedule for milk pick-up during the week. If you are especially interested to get your milk on Wednesdays and/or

Saturdays or any other day in particular, please contact him immediately by telephone.

For vegetable shopping we ask that you bring your own bags, baskets and containers. Any extra paper bags would also be appreciated. Please remember that over 90% of all vegetables are still growing in the fields, so the supply and variety will be small at first and grow as we move toward August and September. Please try to read the bulletin board to keep yourself informed as to what is coming up, what extra help we might need, when, etc. Hopefully by next week we will also have the telephone connected so that you can inform yourself as to what is available before coming to shop.

Thank you to all who have already submitted their questionnaires and July contributions on time. Anyone who has not sent in their July check as of yet is urged to do so as soon as possible. It is extremely helpful to have the monthly contribution in by the 5th of the month at the latest. We are pleased to report that we have at last count approximately $2200 coming in per month in contributions. This represents the support of 36 families. At this point we feel that we must close ranks, in order that the current families receive a reasonable amount of vegetables. If you know someone who was thinking of joining the group this year, please explain the situation to them and say that we hope to be able to serve perhaps a few more families next year.

Please watch the shop bulletin board for the date of our next meeting. And we again ask you to please remember that the key words for this year are patience, tolerance and good humor. Let's try to make things run as smoothly as possible, together!

Our first year started on the 1st day of July 1986. July was chosen because we could be sure that by then a great variety of vegetables would be available. And so it was. We were able to provide in the first year our 36 households with a large selection of vegetables, many of them were stored and lasted into and through the winter as well as blueberries, apples, cider and milk.

Having started in 1986/87 with 36 households we grew rapidly in our second year to 65 households and a budget volume of $55,000. Another service of expansion became the taking on of apprentices who could make a hands-on study of bio-dynamic agriculture, while providing us with much-needed help in our work.

New members have to become acquainted with the basic ideas of this endeavor as we ourselves have to answer again and again the question of the motivation out of which we work together with a farm in this way. The questions go for the social-economic reason in our way of creating a community-supported farm and for the basic farming concept we have. Two articles out of our newsletter are aiming for answers. First the social-economic background.

The healthy maintenance of the human physical organism is the first requirement of any conscious human existence. Therefore the original production of life-giving food must always go on, even when this activity no longer makes an economic profit. We have to produce such food for our spiritual, mental and natural existence: "For any price" or "At all costs." If the price in selling agricultural goods does not meet the production costs, the financial reward of their processing and the income of all other economic endeavors have to be used to meet these costs.

Today original agricultural production is conducted solely under private, profit-oriented aspects. Under these conditions the will for a life-giving quality for these products is hardly or not at all a motive for the producer. In the larger economic field, original agricultural production has a lower or perhaps an equally valued position as industrial production, but never is it given the priority of being seen as a precondition of any industrial production.

The consequence of this economic behavior is a perpetual lowering of the life-giving quality of our food in the last decades. This has been observable for many years but not everyone has consciously taken notice of it. Since the nuclear accident of Tschernobyl, everyone knows that we have to live with qualitatively inferior food as a direct consequence of our economic life. If we want to maintain a healthy physical organism — "if we want to survive" — we first have to try to eradicate the negative consequences of our economy on our food production, and in the long term we have to produce foods that improve our life forces and that do not (or as little as possible) contain substances that damage human life.

1. The agricultural method which, since its founding by Rudolf

Steiner in 1923 has aimed its scientific and practical application towards producing food with life-giving quality as a priority over all other aims, is the anthroposophical agriculture known as the biodynamic method.

2. The realization of original agricultural production has to be rescued from all private and public utilitarian thinking. It has to be understood and undertaken as self-supporting, self-supply of any given community with high quality foods with the help and the cooperation of every single person.

3. This effort has to become everyone's task, without reference to his or her national, professional or social position. Everyone should connect him or herself financially and if desired with advice and practical help with a concrete, specific farm. And everyone who does this should be ready to cooperate with all others who also want this. In this way slowly a sufficient amount of high quality food could be produced at a price that meets the cost of its production.

The above-described cooperation of all is the associative economy as Rudolf Steiner described it. Its foundation and primary task is the above characterized anthroposophical agriculture out of which all other associative economy ("business") can be developed for its support.

—Wilhelm Ernst Barkhoff in an article for the "Bank Spiegel"

The farming concept itself is described in an article by Dr. Nicolaus Romer printed in our newsletter in July 1987:

THE BIO-DYNAMIC FARM CONCEPT — a solution to protect nature; protecting nature with the help of the concept of the closed organism of the farm's individuality.

In 1924, after two years of preparation, sixty farmers brought their questions to Dr. Rudolf Steiner to obtain his advice on conserving the fertility of the earth and to bring a diseased farm situation into a healthy condition. In Koberwitz, near Breslau (now Poland), he presented the idea of the closed agricultural organism and gave them much practical advice.

Now at the end of the century, the preservation of the fertility of the soils, the plants, and the domestic animals is very important, as the diversity of flora and fauna is far more endangered than in

1924. Already in the last fifteen years, 120 species of mammals, 150 bird species, and more than 1000 micro-organisms have become extinct.

In the meantime, farmers on all continents have had practical experience with the ideas and advice from Koberwitz. These experiences have led to an improved practical approach in agriculture, horticulture, and forestry. The following shows some of the basic ideas and methods that are necessary. They show that this farming concept is a cultural task that has to be carried by both farmers and non-farmers.

1. Animal feeding that is in accord with the character of each species as the basis of soil fertility:

The care and keeping of domestic animals — from conception to use — affects their health, their fertility, and their performance.

In a manifold farm organism they find a sheltered environment. They promote each other and stimulate soil-life for a growing soil fertility.

2. To manure means to enliven the soil.

To manure does not just mean to provide nutrients for the soil, it means to animate the life processes of the soil.

Depending on the character of the soil and the amount and quality of forage, we shape the numerical relation of domestic animal species so that we achieve the necessary manure and the substances and forces of the farm itself. The transformation of animal excrements and plant substances into manure is supported by the biodynamic preparations that are derived from specific medicinal herb preparations.

The handling of the manure is divided into two polarities. One is a more conserving phase — one is a phase of fermentation in which the manure is transformed and prepared to serve the soil life.

3. Care of the soil

Only life-filled soil is fertile of itself. Active soil is an organism that has a breathing top cover as a sort of skin, following this downward, there is a humus zone (layer) as a transition zone to the subsoil, which allows the plant-roots to penetrate the subsoil.

In organic soil care we have to prevent a shifting of these layers.

a. In smaller plots, soil care can be restricted to working with the hoe and hand cultivator.

b. In a developed rotation we can succeed in preparing the soil with cultivators that air horizontally and Scandinavian rotary spade harrows — both leaving the soil layers undisturbed.

c. Skim ploughing, field dragging (leveling) and cultivating in big fields have, for a long time, been a good soil preparation for winter wheat and winter rye in combination with a veil of animal manure compost.

d. Deep airing (cultivation) and deep ploughing with land pucker (compaction roller) is tolerated to prepare a seed bed in soils with a deep developed humus layer and natural fertility — if the soil is dry enough.

e. Instead of mechanical energy expenditure one should animate the plant roots with root-stimulating humus substances. The roots of grasses and legumes then become the best helpers of tillage. For that purpose humus substances have to be worked into the uppermost layer of the soil.

4. The plant develops the soil:

Plant communities preserve soil and landscape in (with the help of) forest, bushland, and grassland.

The plant develops on its own deterioration if there is a soil-related community of grains and grasses, cruciferous plants (rape, oilradish, brassica) and leguminous plants (alfalfa and clover species). These plant families are supported in their tasks by medicinal herbs of the families labiatae, compositae, umbelliferae that we again use for preparations for the animals and for the fermentation of the animal and plant manures (compost).

Taking certain plants into the rotation often leads to great improvement of the farm organism. Phacelia and sweetclover (Bokhara) against nematodes; alfalfa and corn for feeding. Fast growing grasses and intercropping enrich the rotation and enlarge the turnover of organic matter in the soil.

The higher turnover of organic matter (carbon matter) raises the fertility of the soil.

Plants placed into the right community incite (animate) each other. We have plant friendships and plant enmity. The plant world — in community — animates itself by its destruction.

The use of hedgerows introduces trace elements into the farm organism. The hedgerows bring birds and insects: the ants, bees, butterflies, bumble bees bring rhythmic order and stimulation into the whole household.

5. Feeding the cow with forages that are grown on the farm for fertility, fodder utilization and longevity.

We can produce 11,000 to 12,000 lbs. of milk per cow a year with a balanced ration of alfalfa hay, hay/grazing, and corn silage.

Besides this amount of milk, the cow produces 40,000 lbs. moist excrement containing 178 lbs. nitrogen, 66 lbs. P2O5, 237 lbs. K2O, and 132 lbs. CaO. With this manure — if we ferment and conserve it rightly — the cow can keep fertile four acres while using two acres for its forage.

6. Rhythmic Order:

Healthy farm organisms connect themselves with the larger environment of the organism of the earth. Embedded in this larger earth-organism and its cosmic surroundings, the growth and the thriving of plants and animals comes to a higher performance with the help of the rhythmic organization of human labor. The farm organism, as a closed organism, develops into an individuality and integrates itself independently into its earthly and cosmic environment.

The concepts that are described in these articles are the ideal background of our farm activity. We try to realize them as far as possible in our concrete situation. We are aware of the fact that one cannot achieve anything worthwhile without a basic concept that becomes an ideal in us, and on the other hand, that ideals can never be fully realized in this world.

The expenses we make on the farm should aim at the ideal concept. I want to include here a concrete budget of our farm — the budget of the year 1988/1989 — to demonstrate the principles that guide us in planning the expenses. (See Budget Example, pages 126 and 127.)

The budget consists of individual budgets of those farm members who realize the farm activities by making expenses. Any expenses in the realm of the farm are caused by men or women on the farm and have to be consciously planned by them. (The result of this planning is the budget.) Matter does not make costs, only people make costs. The expense of electricity for lighting the barn is not caused by the barn or by the animals. It is caused by those who are responsible for the barn and switch on the light. In consequence of this fact, the budget consists only of personal costs. Every expense is caused and has to be attributed to a responsible person.

The budget shows that we finance in the Community Farm only the operational costs. Land cannot be financed in this way. The livestock is an integral part of the farm organism. Once it is there, it reproduces itself sufficiently. Like the land itself, it need not be privately owned. It is not a commodity as long as it is connected to the farm. A practical approach to the livestock question is found in the following:

The community farm is growing rapidly in numbers this year. We will soon need to increase our milk production. Two or three more cows are needed before July comes around. Our heifers have another year and a half to go before they are ready to give milk.

Who is willing to buy these cows? An average Jersey cow can be bought for $325-400. The Community Farm will carry the expenses for caring for the cows. It cannot come up with the money initially to buy them. We can carry operational costs but no capital costs together. We need individual people to provide the capital for this purchase (and to own the animals that come into our use). The purchase can also be made as a tax-deductible donation to the Lukas Foundation if this is preferred.

Once the necessary number of animals is there, the herd will reproduce itself and grow without further capital expenses. Only the services of a bull from outside will be needed occasionally. The money for his purchase will usually be found by the old bull going into beef or by another older animal being sold. Our present bull is provided by the Milford Rotary Club via the Lukas Foundation. Who will provide us with one of the three needed cows? Or with them all?

If anyone would like to give us a cow that I would choose for the herd, please contact me as soon as possible. The earlier we start looking, the better the choice and the price will be.

—Lincoln Geiger

TEMPLE-WILTON COMMUNITY FARM
PROPOSED BUDGET: JULY 1ST 1988 - JUNE 30TH 1989

	LINCOLN GEIGER	TRAUGER GROH	RON LUCAS ED HALL	ANTHONY GRAHAM	JEFFREY SEXTON	DONAT BAY	TOTAL
FIELDS							
Manure		100					100
Mineral	400		350				750
Seeds	340	300	250	300			1190
Preparations							
Plant Protection							
Tools	400	200	100				700
Contract							
	1140	600	700	300			2740
ANIMALS							
Hay	200		300				500
Cattle Feed	4000	300	100				4400
Milk Production	300						300
Pigs			500				500
Sheep		100	100		300		500
Chickens		500	610			500	1610
Fencing						300	300
Bedding							
Vet	200	150	50		50		450
Tools	100						100
	4800	1050	1660		350	800	8660
MACHINES							
Repair	1500	700	450				2650
Parts	4000						4000
Depreciation	2000	2000	1000				5000
Maintenance	200	200					400
Fuel	300	500	125				925
Insurance	300	300	100				700
Transport							
Hire							
	8300	3700	1675				13675

TEMPLE-WILTON COMMUNITY FARM
PROPOSED BUDGET: JULY 1ST 1988 - JUNE 30TH 1989 (Continued)

	LINCOLN GEIGER	TRAUGER GROH	RON LUCAS ED HALL	ANTHONY GRAHAM	JEFFREY SEXTON	DONAT BAY	MARTIN NOVOM	BRUCE KANTNER	TOTAL
ORCHARD									
Trees & Seed									
Sprays		200							200
Manure		300							300
Tools									
Contract									
		500							500
PERSONAL									
Household	18000	4800	2000	4500					29300
Apprentices	4800	4800					50		9650
	22800	9600	2000	4500			50		38950
FARM BUILDINGS									
Maintenance			150	500					650
Repair	600								600
Depreciation									
Insurance									
Utilities	750	240							990
	1350	240	150	500					2240
ADMINIS-TRATION									
Administration				600					600
Taxes	150	850	350		650			273	2273
Dues									
	150	850	350	600	650			273	2873
MISCELLA-NEOUS				200					200
TOTAL	38540	16540	6535	6100	1000	800	50	273	69838

Difficult problems are connected to the question of capital goods like farm machinery and farm buildings. Here the depreciation and eventually some interest can be part of the operational budget. The machines cannot and should not be owned by the community — that in our case is not a legal entity anyhow. We made it a rule that machines are owned by those who paid for them. And they can rightly claim depreciation and interest from the user, something that can be seen as a lease. The machine situation was described to the members in May 1988.

We need machinery in the Community Farm because we do not want to do all the work by hand. We could do it without, since we are 60 families, some 120 adults. We could do most of the work by hand even beside our other professions and tasks. But we have — all of us — a different lifestyle than that, so we use farm machinery and have to carry the costs out of non-farm income.

A typical farm needs a loader or tractor with a bucket to fill the manure spreader; a spreader to bring the manure to the fields; tillage equipment such as a plow and harrow; hay equipment, like cutterbar, hay tedder and rake; and tractors to pull all this. A typical value of such an assortment of mixed age and condition could be $50,000 to $70,000.

The Community Farm has decided from early on, with good reason, not to own machines collectively but to use the machines and to carry the operating costs. These are maintenance, repair and parts and what we rub off through use — the so-called depreciation. Who can own such equipment then? Any individual or group inside or outside the Community Farm that can afford to pay for it and who is willing to allow it to be used by the Community Farm members. Currently the machinery is owned by the Lukas Foundation, Alice and Trauger Groh, and Dr. Paul Corcoran.

Currently we are faced with two problems. First the existing equipment is not enough and some pieces need replacement. Second, in the first two budgets very little has been put aside for depreciation, payments back to the owners of the equipment. For the 87/88 farm budget $5,000 was proposed but not enough was raised to allow this. Therefore, additional and replacement equipment cannot be expected from this group of three alone. (It is worth noting that this dilemma is found on other farms today. The farmers often are forced to "eat up" the depreciation of the machines and buildings. When expenses and repairs are called for, they cover it by going further into debt. A way we do not wish to go.)

What is required now in additional and replacement equipment?

Hay cutter bar	$ 2,000
Hay tedder	1,200
Hay rake	1,000
Transplanting machine	700
Tractor or loader	12,000
Potato planter	500
Plow	600
Total	$18,000

We have already found a donation to cover $1,700, equal to the cost of the hay rake and a transplanter. That leaves $16,300.

There are two possible ways to come up with this urgently needed money.

1. An individual or family makes a tax deductible donation to the Lukas Foundation designated for the purpose of equipment purchase.

2. An individual or family can purchase directly the needed equipment and make it available to the Community Farm, knowing it will be used and kept in good repair.

It is worth noting that individuals can also join forces, either as individuals or as is done elsewhere, as a corporation which exists to purchase larger pieces of machinery. In this case a more formal arrangement is possible where the Community Farm could lease the equipment from the corporation, for the cost of depreciation and interest and perhaps the administrative costs. So what does this mean? Let's say you had $75 a month of income which could be useful to you as a donation. Twelve months at $75 is equal to $900. That would purchase the needed potato planter ($500) and allow $400 to cover that portion of another needed piece. Donations can be reached on either a spontaneous basis, or in cooperation with us. Certainly, a commitment to give x per month for so many months allows, say the Lukas Foundation, the equipment to be acquired now, when it is needed, secure in the knowledge that the funds are pledged to back up its purchase. Conversely, say a member or friend wishes to purchase a piece of equipment for his ownership, perhaps even for his own eventual use. Consultation with the Community Farm active farmers could direct this friend to exactly the type, size, etc. which would be our preference. Direct use by the Community Farm would be accompanied by responsible care and use, normal repairs and the necessary depreciation.

We are ready to entertain any additional ideas or suggestions. We urgently need to update and expand our machinery.

When the farm expands by using more land and having more "farmers" (members), the question of investment capital naturally arises. This question keeps us engaged. We work on it without having final solutions. The wrestling with these questions in the pursuit of the farm shines up in the following articles of our newsletter:

FOOD FOR THOUGHT

At our last meeting in Greenfield, Anthony Graham gave a report on the financial situation of the Community Farm. He showed that the money that comes in from the members in support of the budget does so far meet the running costs of the farm.

By sharing the costs of the farm, the members of the Community Farm become, in cooperation with the other members, entrepreneurs of the farm. They need not distribute the costs of the farm operation on the products they consume. Only the small amount of goods that go on the market outside the community carry their part of the costs — they are sold.

Having lived a few months with this new approach to farming, in which individuals come together to make a farm possible that has no private profit motivation, but only the aim to serve the community with pure life-giving food, we can see already that this is a viable new way to run a farm.

More households than in the first year show their interest to become members of the Community Farm. They want to enter the enterprise with the new agricultural year beginning on the first of July 1987. The active farmers have to deal with the question of how many families can be served with milk and produce next year. There is certainly a limit to the number of families the Community Farm can reasonably serve.

In no way should we encourage families to apply for membership before we have answers to some basic questions. Counting those who have applied so far, we would have more than fifty member families in the coming year compared with 34 in the first year. To serve more households than before we need more human labor to be put into farm work. And we need more usable land, better tools (machines), more animals to bring life to the soil with their manure, and housing for those animals.

More human labor will be available if more of the members on a regular basis work on their own land and/or if some are ready to house, teach and introduce to their social life apprentices and give them a certain amount of money. More land will be available if we can clear, stump, and fence the bush-land we already have. That means capital investment. Capital investment is needed for more machinery, for more animals, and for farm buildings. The cost of farm building can be lowered dramatically if we come together in a "barn-raising" or do the necessary alterations in the existing buildings in that fashion.

Although we have a working procedure to cover the operational costs, we have no procedure yet to raise the necessary capital. In the first year two members and the Lukas Foundation invested in machinery and fencing that serves the Community Farm. This will not be possible in the future.

The Lukas Foundation is about to buy Echo Farm. Through this, the farm land that goes with it can be insured for our purpose for the future. The present owner could no longer carry the heavy mortgage that was on the land. This land would be lost for our farm if the Lukas Foundation were not willing to buy it. In order to make the financing of this bargain possible — a bargain that should take this farmland finally off the market — we have to draw on the help of more and more people — especially those who have an understanding of the necessity to preserve this farmland.

All this shows that every member and certainly every new member should contemplate how he individually and/or in cooperation with other members can contribute to the investment capital (or to help lower the capital cost). In this way, every year more and more farmland can be taken into the care of groups that without any profit motivation want to produce life-giving food and want to heal the organism "earth" at large with the concept of Anthroposophic (B-D) farming.

WHAT MONEY ARE WE DEALING WITH?

At a farmers meeting in Harlemville, N.Y. in January we identified three different qualities of money coming from different sources that we need to establish and to keep up healthy farm organisms.

First of all we need to buy the land and/or free it permanently from mortgages. This can only be done by donated money and it has to be done finally. The land has to feed, house, and warm us; not to give profit to investors. Where do we stand with our community farm in this regard? Ploughshare Farm had been heavily mortgaged by former proprietors. This mortgage has been moved on to the houselots of those of us who live up there and want to develop the land. I see this as a heavy and unjust burden that we, who use the land together, have to have in mind. Echo Farm had a terribly high mortgage when it was taken over by the Lukas Foundation. This mortgage has to be paid off bit by bit by donations of friends. We who use the land have to realize this fact.

A different sort of money is needed for investments on the farm — farm buildings and farm machinery — here credit could come in and is needed. If someone wants to buy a needed machine for his farm and does not have the necessary savings, he or she should be able to get credit. The repayment of this credit and the interest can go into running costs of the farm. The farmer can lease the machine to the farming families. The principle here simply could be: Farm machinery is owned by those persons or corporations that pay for them. It is used by those who need it. The users come up for the cost the machine holders have through this ownership (depreciation and eventually interest). Long experience in farming has shown me that the farm operation can never come up for more than three percent interest. This is because farm machinery is used only sporadically. Machines in industry are used continuously. We see here that we have to create in the future a new form of farm credit.

The situation in the TWCF is the following: We were not able to bring our 87/88 budget to a level where depreciation could be paid to the owners of our farm machinery (estimated $5000). This diminishes the ability of the machine owners to buy new machines. Ergo: We are using up our machines without replacement. This is a typical situation in today's farm scene.

The third form of money we need is the money to cover the running costs of the farm. This can be called purchase money. For the running cost we can and should neither use donations nor credit. Here our households use the money they otherwise would spend in the market, in shops, etc. They can spend it in relation to what they have available, what the running of a biodynamic farm is worth to them, and what the farm needs. What the farm needs in money to cover its costs is highly influenced by the behavior of all participants: Whether they come together in times of great need in labor to solve the problems manually or whether they buy costly specialized machines for that purpose. Whether they care for the machines they have and repair them. Whether they buy lots of food outside their farm and reduce their ability to spend enough cash on their farm, or whether they make good use of their own farm with the help of canning and freezing. The running costs can be influenced by all of us.

We are often asked how the decision-making process occurs on a farm with 75 member households and with a group of active farmers of about 4. We have an active farmers' meeting once a week. This meeting includes the apprentices and covers the week-to-week decisions and information. It is open to all farm members too. Only a few can make use of it. It has to be said here that never have members of the larger farm community — that means the non-active farmers — tried to influence the farm decisions of the active farmers. Besides these work meetings there are a few membership meetings. One is dedicated to the presentation and discussion of the budget; one is dealing with financing the budget through pledges. Some meetings are connected to farm work and serve to inform the members about what is going on. Some meetings have a more social character (Mayday, Thanksgiving, Harvest Festival).

A view into a members' meeting gives the following report from February 1987.

SPECIAL ALL-DAY WORK MEETING

On Saturday, January 10th the Community Farm held a working session beginning at 9 am and ending at 5 pm. Fifteen people — sometimes more, sometimes less — took part. The evening follow-up meeting, meant as a presentation of the topics and the results worked

*upon during the day, was cancelled because of the heavy snowfall.
It took place on Saturday, January 17th instead, with more than 40
people attending.*

*The meetings aimed at bringing to mind and discussing the full
scale of our farming effort viewed under the aspects of:*

1. What do we imagine the farm to be like in 1995?

2. Where are we now at the change of the year?

3. What are we going to do next year to achieve our goals?

*After an introduction and under the leadership of Martin Novom,
we discussed the organizational aspect of our Community Farm.
Lincoln Geiger and Ron Lucas reported on the land and the animals,
Trauger Groh on the crops and the labor situation, Jeffrey Sexton
on the buildings, and Anthony Graham on the finances. The reports
you read here are what the participants managed to write down about
the meeting and their contributions. They are the beginning of half-
year reports that we plan to give regularly in the future. The winter
reports could deal more with the long-term and mid-year planning;
the summer report more with the old and the new budget.*

*Some figures on the land first. The Community Farm uses land
in 5 different places. Approximate figures of what is in use now:*

Echo and Crossroad Farms	8 acres
Temple Road Farm	7 acres
Plowshare Farm	9 acres
Temple Gardens	2 acres
Orchard	7 acres
Total	33 acres

*We intend to clear, stump and plant 30 to 40 more acres by 1990.
So by then we could be farming 60 acres. That leaves then 148 acres
in forest and bush and the possibility for more clearing. Projections
on the future of the farm are based on the figure of 60 acres. It
became very clear during the discussions that the future will be
destined by the question whether Plowshare Farm with its larger
acreage can be taken more into farming so that it will again in the
future have its own cattle herd, and whether other land for long-term
use is coming up. Included in the projections is the fact that we can
use hayfields on the Kantner's Derbyshire Farm.*

FARM ORGANIZATION

With the presentation on organization, I began the day with a brief

134

look at what we might be like in 1995. We will certainly be a larger group. How large is hard to imagine. Certainly more members and perhaps more active farmers. To get a perspective on this, I can look to just how much can happen in 8 years. My 3-year-old daughter would be 11 years old. We will have had over 400 Monday active farmers meetings! A lot can happen in that space of time. My inclination is to look to how much we can accomplish.

For 1986, the TWCF had a wonderful birth. Those of us who were with it from the beginning had that beautiful birth experience. It was difficult and precarious. Had we not had the goodwill of so many, it might not have happened, or perhaps it might have been a case of "crib death" early on.

We had some very valuable monthly members meetings, both in working out details and in creating a sense of bonding as a group. We had several warm social gatherings and best of all, a terrific Michaelmas festival that we look forward to creating again in 1987.

In 1987 as regards organization I see new members helping to create a different configuration of people than we have worked with in 1986. We will have many, if not all, of our original group together again for the 87-88 season. However, we need to recognize that we will have a fairly large number of new people who come "on board" with perhaps a different set of views. They come to something already formed. They see it in a new way, while we may see it as already existing and part of our lives. I won't dwell on this, but I feel it is important that the older members make a special effort toward helping the new ones. We are a group with enormous good will, so I don't foresee a major problem. We need only to reach out and welcome them in the best sense of the word.

It is worth speaking about the nature of the Community Farm. What is it? If you have ever tried to share your experience of it with someone who doesn't know about us, then you will recognize this dilemma. How do you describe such a thing? I've often heard it said "well, it's a kind of co-op." I think that is a misnomer, and I'll tell you why. I've been in several co-ops so I speak with a bit of experience. The food co-ops I've been in:
- require no lengthy commitment
- focus around getting food cheaply
- do not have a commitment to obtaining food of only the highest

nutritional quality. Often there is a mix of members with a cheap food vs. organic food division among them.

• *do not have a commitment to provide financial support for the source of the food. In fact, it usually comes, as in the supermarket, from a faceless warehouse and a faceless grower. (The Northeast Co-op Warehouse in Cambridge, Mass. is a welcome exception. They do help a lot of people connect.)*

I now contrast that with what I think the Community Farm provides:

• *a commitment to a specific local farm activity*

• *a desire for only food grown biodynamically and therefore hopefully with the greatest amount of life-giving forces*

• *an intention to promote a sane and stable life for the farmers by providing economic and cultural support*

• *personal financial commitments by the members for one entire year.*

Add to this that the Community Farm owns nothing. All land, equipment, buildings and animals are owned by the members individually.

Finally, the way in which we organize ourselves is quite different. We create a really great opportunity to build trust and social strength by our method of assigning financial responsibility and allocation of the food. Because we are very flexible as to what regular amount families contribute monthly, it requires an engagement on the part of members. One cannot long be passive when you are told, "well, contribute what amount you feel is correct for your household budget."

And with the distribution of food, except for milk, this openness to engaging the social process continues. Milk is arranged ahead of time between Lincoln and individual households. There is a finite and fairly predictable amount of milk. With our produce, we say "the food is a gift of nature and since you as members have covered the costs of producing it, take what you need."

All of this may give you an idea why I think we are not just a co-op.

—Martin Novom

Out of this meeting again comes this article on the Farm Organization. The organizational aspect is clearly worked out in another statement in the Newsletter about it.

ABOUT THE ORGANIZATION OF THE TEMPLE-WILTON COMMUNITY FARM

The Temple Wilton Community Farm is the free association of free individuals with the aim to make a farm possible that provides life-giving food for the local community and conserves the natural environment. The members are economically organized in households; out of their household income they cover individually and together the operational costs of the farm. They are legally not connected and have, therefore, no legal claims on each other. So, if a member does not do the work on the farm he promises to do,

- *if a member does not pay the share of the farm cost he delcared he should pay,*
- *if a member harvests more produce for his household than is socially responsible,*
- *if a member — against his intentions — does not come to meetings to discuss his and the others needs in community,*
- *if a member works on the farm without first coming to an understanding with the other farmers,*
- *in short, if any one of us goes against his own or her own expressed will and intention, the others have no claim against him or her. The only thing that the others can do in these cases is: to jump in for the others to substitute for an eventual loss.*

Everything concerning the farm derives out of the constantly renewed free will of the participants.

A topic that arises again and again is the question of how we provide the necessary labor for this diversified farm organism. One thing we agreed upon: We do not substitute the financial pledge toward the budget by labor input. The sharing of the costs and the bringing in of labor are seen as two separate things. We can rely on certain members for help in case of need. Other members are not in the position to work on the farm. Our experience showed that the members respond better to emergency calling than to regular workdays. It is important to provide reasonable labor for anyone who wants it, while at the same time never to put blame on those who cannot take up this sort of work. In the newsletter we find the following article on labor:

Farming, like industry, needs long-term planning. This is different from all craft or "household" activities like cooking, processing, and many social services that can be done more or less spontaneously. If we think that a plant rotation on a field often takes seven years, a forest generation is more than 100 years, that it needs at least a generation to build up a good herd, we realize how well planned farming has to be. Therefore, a farm needs long-term commitment of at least one family, or better yet, a group of dedicated people. Besides that, it is necessary that many people again and again share the work — for recreational reasons, for educational reasons and for the benefit of the place.

In a very abstract way the labor input in a farm is measured in full-time laborers per 100 hectares (160 acres). A full-time laborer would be someone who works 50 hours a week on the farm. The average labor input in agriculture in Germany today is three full-time laborers per 260 acres (100 hectares). The records of biodynamic farms in North Germany show that there we find eight full-time laborers per 260 acres. Naturally, the smaller a farm is, the higher the labor input. If we scratch together all hours that were spent on farm work on the Community Farm in the first 6 months, we would perhaps find a labor input of two full-time laborers on 26 acres or 20 on 260 acres. One reason for this high input is that we are mainly gardening, not farming. This 100 hours per week (2x50 hrs) that are spent working on our farm — it is even less if we take into account the long winter season — could be raised if out of 50 participating families one member of each family would spend 2 hours per week on the farm. If we think that we are not very skilled, we would perhaps need 4 hours per week from 50 people. For many reasons, of course, this will not be possible — still it is good to have this time-labor relation in mind.

It could be very good if each of us would use daily labor reports to find out how much time we spend on farming (our daily bread), how much on householding (including the crafts and social services, education), how much we use for ourselves (recreation, art, self-education), and how much for work in industry or business. Doing this we would come to a greater social awareness.

What is the labor situation on the Community Farm in 1987? We know that the Lukas Foundation has purchased Echo Farm. In consequence of this, Lincoln Geiger and his family will have to build themselves a new house on the other side of the road. The living space for Glynn and Anthony Graham will be totally reconstructed in the old farm house. This means that two people who have had high labor input in the Community Farm will next year be burdened with building activities and therefore not be able to do as much on the farm. At the same time there are more animals to be looked after, there are bigger crops to be grown for the larger number of member-households, and the clearing, stumping, and fencing process on the new land will need much labor as well.

What can we do to meet these demands? Two things are possible and necessary. A growing input of occasional (or regular) labor by Community Farm members and the hiring of one or two apprentices if they become available. For every farm apprentice we will need an extra $200 per month in our budget — if we find boarding in one of our homes.

Can we make the effort to bring in more labor ourselves as members of the Farm Community? Any suggestions for how to organize this in the best way are very welcome.

Sometimes we come into the situation that a crop fails. Besides the mistakes that we make, this is due to the rundown condition of our soils that have to be built up in years to come by careful crop rotation and well-conditioned manure. For the moment it is necessary that everyone participating in the farm can understand why certain crops have failed and are not available. Out of this situation we can see the importance of the following article:

OUR POTATOES

To work a new field that has not been in proper use for decades — as we do with our vegetable field on the Temple Road — is an adventure and bears grave risks. This vegetable field has been a hayfield for so many years without proper manure, just run on chemical fertilizer. The poor grasses and mosses, the ferns and the poison ivy that grew in the orchard part of that field showed the state of degradation of that soil. It seemed like a wonder that the front part that came into use last year gave some good crops. Besides the

139

various efforts from the farmers, this was mainly due to the ample rainfall of 1986. The moisture helped to break down the old sods of the hayfield that were in the ground, so that beginning with mid-July suddenly the crops started growing.

This year the second half of the hayfield, the old orchard part, came into use. That part that is higher, drier and poorer in soil. Early in the spring we started working, preparing the soil, helping it with a lot of cow manure from Greeley's dairy farm. We prepared it in order to plant potatoes, rutabagas and fodder kale. We bought the best planters we could get and the potatoes came up nicely. Then came the bugs: Colorado beetles with their nice yellow and black stripes. We spent every second day about 2 hours picking bugs. We did not want their larvae to eat all the leaves. For a long time we kept them in good control. But there was not enough rain. The soil was mostly dead dry. The old sods, unable to break down for the lack of water, lay on top and beside the seed potatoes like a carpet keeping the little rain that came from reaching the roots. The available humidity was just enough to let the visible plant grow but not enough to make tubers. We had set up an irrigation system to help a bit with the water, but the two wells of the farm could not give water for more than 4 hours irrigation per week; 24 hours irrigation per week could have helped. Two weeks ago we cut down the rest of the greens. It was not much anyway. The beetles had, in the end, multiplied and eaten it all up.

We started digging and bringing some potatoes to the shop. Where 8 middle-sized potatoes should have been per plant we found 2 to 3 small ones. It takes nearly one manhour to dig one bushel! The early variety we dig now is very good in taste. With its yellow solid flesh it resembles continental European salad potato varieties. Anyway there will be very few potatoes this season. We will have two bushels twice a week throughout September and October and nothing for the rest of the winter. This is a hardship for all members of the community farm. We are trying to get winter potatoes from another B.D. or organic farm. Since in winter we will have to buy potatoes in any case, should we not try to put good ones in our shop?

I am sure that next year we will be able to have better results on this field, since the now dry sods will be broken down under the condition that we can make water available for our irrigation.

Rutabaga and fodder kale are sort of hanging around on that field not knowing whether to die (some do) or to wait for wetter days. If they survive, they can grow late in the year. They like to grow when the nights get longer.

Very practical questions come up with certain items, questions of production and of the necessary recycling. Foremost here, an understanding of the facts has to take place. This shows up for eggs and chickens:

ON CHICKENS, EGGS AND EGGSHELLS

The Community Farm members need eggs . . . eggs of chickens that are free to run around, that eat grass and insects, worms and weeds. If you keep chickens indoors on mainly a grain/corn diet, you breed cannibalism; they start to eat each other's feathers. So we want to graze our chickens and give them plenty of space. But, they are threatened by wildlife, by racoons and foxes. So we have to fence the chickens in. If you fence just one area you will find that soon that area is bare of grass. So we will need three fenced plots for a flock of chickens that can be rotated. We now have chickens in Greenfield and at Temple Road Farm. We think we should be able to provide 2 eggs for every member of a Community Farm household per week. This makes it necessary to keep about 80-100 laying hens.

The calcium that the chickens build up in their eggshells is of great value and should never be thrown away. In every kitchen there should be a bag or box to collect eggshells. If you compost yourself, add them, broken into little pieces, into the compost. Or recycle them by bringing them to the farm shop and into our common compost. We can make a very precious Biodynamic preparation from them known as Barrel Compost.

What to feed the chickens: Since we have no grain ourselves, we have to buy chicken mash and cracked corn for them. When they have access to grass in the growing season, this input can be reduced. Chickens are grateful for all kitchen refuse. The Novoms started to bring their kitchen scraps to our farm once a week in tightly closed plastic buckets. If ten to twenty families would do so, we could feed the chickens mainly on that refuse. That would be of great advantage for the quality of eggs and for our budget. Our farm can make good use of any organic stuff that leaves the households.

And on another instance: To make a laying hen out of a freshly hatched chicken takes six to seven months. They could live and lay eggs then for five to seven years having a molting period every season where they do not lay. In modern egg production a hen is allowed to lay only one season. We keep them for two seasons; keeping them longer would mean less and less eggs every season and tougher chicken in the end.

On the first of October we sent 38 chickens to the butcher to make room for the next batch of 38 that we have raised since March. Some of the 38 were still laying and the new ones are just beginning to lay. We found the first egg on the second of October. So abruptly there will be even fewer eggs in the shop and toward the end of October we will come to normal again. This refers only to the chickens at Alice and Trauger's place. There are some more chickens in Plowshare Farm and in the place where the Bay's lived on 101 before they moved to Plowshare Farm. These flocks were not very productive this year due to the fact that they moved from hand to hand and due to old age. So we have a real problem in supplying enough eggs. This problem will continue for a while.

The 38 butchered chickens have arrived at the shop. They are not for frying, they are for boiling, to make chicken soup and fricassee. Boil them until the meat falls easily from the bones and you have top quality chicken meat.

To make our farm organism richer and more complete we need more arable land to grow small grains and fodder crops and to include our vegetables into a more balanced crop rotation. This direction is pictured in the following contribution that describes one first step in this direction, a step that has been followed by others. One of them is the purchase of a second-hand combined harvester that will make grain growing more feasible.

Since people in and around Wilton came together to make new farms possible in a time when farming has come to an end in New England and one farm after the other is given up; since that year 1986 they are thinking of reintroducing the growing of grain into the area. Grain is the most basic food we have: "The daily bread." What makes grain in the form of rye, wheat, barley, and oats so specific and important for nutrition is its relationship to the forces of light and warmth. They set their fruits on high stems exposed to light and wind, not shying away from it like the dark fellow, the potato, that becomes green and with that it is poisonous when exposed to the light. The most light related of the grains is rye. He has the highest stem. He "wants to see the light" when being sown which means he does not want to be covered by much earth. He grows best when the light can play around him all the time. That means that the stems should not be too close to each other.

So, grain has to come to our farm soon. It does not come easy. Only a few fields we have are arable, and we lack all facilities for cutting and threshing the grain. Besides a little rye grown for bedding our animals, a two-acre oat field was the first grain field in the area. It was a heart-warming process for me to prepare the field and to sow the oats by hand. I had done this thirty years ago, taking — as again now — the rhythm for the work from a poetic sower's song. The oats were meant to prepare an old hay field into a vegetable field for next year. Oat was chosen because of his ability to deal with the raw undigested grass sods of the plowed hayfield. The oats grew up golden and beautiful. How could we harvest it? No binder is around here available; no combined harvester; no man or woman with the skill and the tools to cut it by hand and bind it. So it was decided to deal with the oats like with hay: to cut it and bale it, thereby losing most of the grain.

One day after the cutting, in the evening, Alice and I went out to collect some of the stems with the ears and bind them into sheaves. One can bind the sheaves using the straw with the ear itself as a string around a bundle. Twelve sheaves we made and brought into the barn. They will decorate our Harvest Festival. Next day a week-long rain started that left the oats as bedding straw. The field is green with new oats now from all the kernels the rain and baler threshed out of it. The question remains: Could we create a situation where we — all the farmers — together would harvest such a field by hand — without machines? It is hardly possible, although if done it could create a deep feeling of joy and satisfaction.

Our Community Farm is not working in such a way that the group of non-active farmers or members — which are the vast majority — is paying in regularly and sufficiently, and the active farmers are free of any financial concern to farm in the way they think is right. In reality the active farmers and their households stand and fight in the same line as all the other members to secure the financial covering of the budget and with it the freedom to farm in an ecologically sound way where nature produces out of his forces healthy, lifegiving food. Every budget year we have to raise the question again: How do we make ends meet this year? The gratifying, satisfying fact is that we active farmers are not alone in this struggle. We draw not only on our own resources but on the ingenuity and resourcefulness of a larger group of people who have taken on the striving for a renewed concept of farming in accord with nature and in favor of the real needs of mankind. Opening up the fourth year of our operation we can say of the financial side of the farm that it was hard to make ends meet, but at the end of every budget year so far everything had been paid, no one owed anyone anything and the farm had developed further towards its ideal.

We can round up this documentation of the Temple-Wilton Community Farms with an article out of the farm newsletter from November 28th, 1989.

If we want to judge rightly the current situation of the Temple-Wilton Community Farm, again and again and from various points of view, we have to look at the fundamental aims that made us and still make us work together. One fundamental aim is: To work together as a group with the land that has been made available in such a way that nature — not we — can supply us with the greatest possible variety of basic foods in the most rational and cost-effective way, and at the same time can regenerate itself to the fullest. This supply and regeneration means practically: To have the needed food for the cooperating households, not less and not more, plus the necessary seeds for the next planting, while protecting and developing the natural environment.

We have to realize that neither the free market economy of the West nor the centrally planned economy of the East have achieved that goal. Under both systems we find an irresponsible destruction of the natural resources: soil, water, and forest. We find the majority of the food of an inferior inner quality (junk food), and we find an insufficient amount of food for a great many people.

Three means of achieving our goals can be mentioned here. The first is to deepen our understanding in a concept of farming that provides sufficient food of high quality without exploiting soil, plant, and domestic animal. This concept is the idea of the farm organism where the harmony of the organism is the basis of the productivity. This leads, over the years, to an ever greater economy in the use of substances and energy. This article is not meant to describe this concept in detail.

The second way to reach our goal is to let as many people as possible and as it is reasonable, participate in the tasks, the labor, the joys, the fruits of the farm. It is not reasonable to have more people participate in the farm in some way or another than the farm could basically feed. Doing this would eventually deny another farm organism of the necessary support.

The third way to develop and carry a farm together is the sharing of the cost of the farm and to cover it by non-farm income. This sharing and covering of the cost can be seen as a permanent task of the Community Farmer. It means that not the people who actively and regularly work the farm alone are confronted with the cost, but many families with them look after the money side. This is something unique in the farm scene of this world that, as in our case, 75 families are actively concerned about the financial side of "their" farm.

In our situation there are again three different ways to deal with the costs of the farm:

1. To activate the farm organism in a way that major costs will be reduced and eventually be obsolete. On our farm one of the major expenditures is the cost of feeds for the animals: for cows, sheep, pigs, and chickens. The necessity to buy so much feed lies in the fact that we have not enough plowland to have various fodder crops. Up to now, our farm has vegetable fields and hayland together with some pasture mainly in undeveloped wood clearings. By this we miss the necessary feed grain and the protein-rich mass producing green crops as alfalfa, red clover, sorghum, and tall grasses. And we miss the benefit for the soil in root residues and nitrogen fixation that these crops bring. To provide more plowland and to introduce a rotation that includes grains and green crops would make the farm more economic and would lessen the need to buy outside.

2. To provide more services by the active farmers that produce household income for them, and lowers thereby their need for income out of the basic farming. With the coming of our own grain, one farmer could start milling and baking for the Community Farm households. Where now 30% of the butter used here comes from our farm, it could be 100%. Other processed goods could arrive out of the surplus production that is not taken raw by the Community Farm members.

If we think that the active farmers cannot work more hours than they do now, we will have to make provisions that sufficient help is on the farm to do all this. We need more regular help through Community Farm members and a housing situation and a training program that is effective for apprentices to spend time on the farm. Besides more help to do all this, we certainly need, again and again, financial help to make the necessary investments: grain mill, baking oven, milk processing implements, and others. In consequence of our shop providing more and more processed goods, our needs would be better served.

3. The third way to meet our costs is the sharing of the budget by all members through free pledges. Since new members entering the Community Farm often ask for a guideline for their pledging, I want to describe a valid concept for this as it is practiced in Brookfield Farm in Amherst, Massachusetts. There, a basic thought in the sharing of the Community Farm costs was to leave out of the consideration all children. Children have no income to share and one can come to the decision that their upbringing should be carried by the larger community, not only by their parents. In consequence of this decision, the farm costs should be divided through the adult members of the community. How would that eventually be for the next budget? If a realistic budget would afford $100,000 and the number of adults in the community were 150 and we would manage to have $20,000 as other income through extra services of the active farmers, it would mean: $80,000 divided by 150 = $533 per adult per year or $44 per adult per month. Having a $90,000 budget and income of $20,000 means $70,000 divided by 150 = $470 annually or $39 monthly. These figures are hypothetical but perhaps not far from reality.

To proceed like this should be a guideline, not a law. If every participant has in mind that if he or she cannot come up with the share derived in the above way, another member has to come up with *a higher sum and if we watch closely the social situation of our neighbors, we should always be able to make it financially.*

(The articles from the Farm Newsletter used here were written by Trauger Groh and by other members of the Temple-Wilton Community Farm.)

APPENDIX B
Resources

ARABLE

It's tough, if not impossible, for small or new producers of food to get credit from banks and other lending institutions — even when such businesses have established track records and solid markets. Since 1984, an organization known as ARABLE has provided credit to farmers and related enterprises in central Oregon. While ARABLE is not an available resource for farmers in other parts of the country, it does provide viable model for alternative financing arrangements. This kind of organization could flourish elsewhere.

ARABLE is a non-profit community investment fund dedicated to the development of a regional food system by keeping local money in local circulation for productive purposes. Membership is open to all individuals and organizations wishing to use their savings to promote local production, distribution, and consumption of food and fiber products.

Through ARABLE, deposits of individuals, businesses and organizations have made it possible for dozens of individuals and groups to build small businesses while enhancing the supply of local agricultural products. Working with a base that has been steadily built to about $390,000 in assets, ARABLE has worked to foster mutual interdependence between producers and consumers, and between rural and urban families, groups and individuals.

To accomplish this, ARABLE pools the human and capital resources of its members, linking producers, related businesses and consumers through its loan programs. The loans are administered by ARABLE in cooperation with a host financial institution. At heart, ARABLE is simply an association of people allowing their savings to serve as the basis of a loan program with specific goals. It is managed according to practices and standards common to other professional lending institutions.

ARABLE maintains a contractual agreement with the Oregon Urban-Rural Credit Union (OUR). This agreement allows ARABLE to administer its loan programs while the Credit Union manages the actual funds in the ARABLE loan programs. The association with

OUR permits ARABLE members to deposit savings into a special ARABLE account in their name. OUR administers the ARABLE loan accounts, provides technical assistance to the ARABLE staff, and disburses and collects all loan funds.

ARABLE has also created a Trust Fund that provides a vehicle for tax-deductible gift money to pass to projects supported by ARABLE to complement the placement of investments. Business recipients of ARABLE loans are encouraged to place a small percentage of their profits in the ARABLE Trust Fund.

To Learn More:
ARABLE
P.O. Box 5230
Eugene, OR 97405
503-485-7630

MONDRAGON CO-OPS

The Seven Stars Farm in Kimberton, Pennsylvania is in the process of forming a Mondragon-style Co-op. This is a relatively new organizational form just now being introduced to the United States by a group called the Trusteeship Institute. But it is an approach to business that may well spread in the U.S. as the traditional family farm continues to be driven out of business and consumers react with displeasure to the model of corporate agribusiness.

Different from the usual cooperative, the Mondragon model is said to represent a sophisticated new economic approach that can be applied not only to industrial but also to agricultural enterprises. Begun in Spain during the 1940's by a remarkable Catholic priest and his youth group, Mondragon Cooperatives are now the top producers of appliances and tools in Spain. There are over 100 organizations patterned on this model, which is neither socialism nor capitalism, but a blend of the strongest aspects of both.

The Mondragon model is built around the realization that there is only one thing that will assure that each worker is as fully invested in a company as an owner — capital. Ownership entails placing some capital at risk. That risk insures that the owner will care deeply about his or her business. So, to insure this in the case of every worker, every member of a Mondragon-style cooperative is required to loan the cooperative a substantial sum, the equivalent of a third of an average year's salary, which in the U.S. would be equal to about $6,000. New members don't have to come up with this capital on day one. They can simply sign a note and it will be withdrawn from their salary over time.

Mondragon Cooperatives institutionalize permanent employment for the worker-owners, financial support for the larger community, active participation in each business by its worker-owners, and efficient management. The Cooperatives board of directors is made up of members of the cooperative.

To translate this form to existing business and farms, The Trusteeship Institute usually recommends that the employees set up a cooperative corporation to buy the original corporation from the owner. In some cases, popular Employee Stock Ownership Plans (ESOP) may be used to convert a traditional corporation into a Mondragon-style worker-owned business.

To learn more:
The Trusteeship Institute
Baker Rd.
Shutesbury, MA 01072
413-253-7500

Sources of General Information on Farm Preservation and Support

The **ALTERNATIVE FARMING SYSTEMS INFORMA-TION CENTER** is part of the National Agricultural Library, the world's largest collection of information on farming and gardening. The Alternative section is set up as a reference service for anyone interested in alternative farming practices, and it can make complimentary computer searches of the vast AGRICOLA database.

AFS Information Center
National Agricultural Library
Room 111
Beltsville, MD 20705
301-344-3704

The **AMERICAN FARMLAND TRUST** is a national, non-profit group primarily dedicated to the protection of the nation's best farmland. Over the years it has responded to hundreds of state and local requests to help select and implement farmland protection programs, often using the methods standard to land conservation trusts: adoption of conservation easements, special districts for agricultural activity, and programs to compensate landowners for voluntarily relinquishing the development rights to their land.

American Farmland Trust
920 N Street NW — Suite 400
Washington, DC 20036
202-659-5170

ATTRA (Appropriate Technology Transfer for Rural Areas) is an organization established by the U.S. Department of Agriculture Extension Service to provide general and technical information on sustainable agriculture. It answers questions and fills requests for specific information from a base of advisors and a large library.

ATTRA
P.O. Box 3657
Fayettsville, AR 72702
800-346-9140

THE BIO-DYNAMIC FARMING AND GARDENING ASSOCIATION is a nonprofit corporation whose task is to advance the principles and practices of bio-dynamic (B-D) agriculture. To this end, the Association supports a quarterly magazine entitled "Biodynamics," publishes books, offers a biodynamic advisory service, supports training programs, sponsors conferences and lectures, funds research projects, and supplies biodynamic preparations.

Bio-Dynamic Farming and Gardening Association
P.O. Box 550
Kimberton, PA 19442
215-935-7797

The **CENTER FOR RURAL AFFAIRS** was formed by rural Nebraskans concerned about the role of public policy in the decline of family farms and rural communities. A non-profit organization, its purpose is to provoke public thought about social, economic and environmental issues affecting rural America. It publishes a newsletter called the Small Farm Advocate, and engages in innovative projects focused on topics such as emerging technologies, sustainable agriculture, rural economic policy, and Rural Community Development.

Center for Rural Affairs
P.O. Box 405
Walthill, NE 68067
402-846-5428

The **COMMUNITY FARMS/CSA PROJECT** was initiated by the Bio-Dynamic Farming & Gardening Association to spread information about community farms, hold conferences, maintain a complete list of all CSA's, publish a newsletter, act as a clearing house for CSA ideas, and provide assistance for those wishing to start a CSA.

Community Farms/CSA Project
c/o Bio-Dynamic Association
P.O. Box 550
Kimberton, PA 19442
215-935-7797

The **DEMETER ASSOCIATION** is a non-profit organization established to certify biodynamic farms and gardens. They maintain a list of certified farms and products in the USA.

Demeter Association
4214 National Ave.
Burbank, CA 91505
818-843-5521

THE FARMLAND PRESERVATION DIRECTORY is a sourcebook of organizations, models, and printed materials oriented toward helping preserve farms and farmland. It is available from the Natural Resources Defense Council.

Farmland Preservation Directory
Natural Resources Defense Council
122 E. 42nd Street New York, NY 10168
212-949-0049

THE GRANGE is the oldest agricultural organization in the nation. Family oriented, the Grange is a fraternal organization which, at the local level, seeks to foster fellowship and to support community projects. At the national level the Grange lobbies Congress and generally seeks to improve the economic well being of families involved in agriculture.

The National Grange
1616 H Street, N.W.
Washington, DC 20006
202-628-3507

The **INDIAN LINE FARM** CSA offers a directory of CSAs and a booklet of budgets and perspectives for starting a CSA. The cost is $10. At the time of writing this book, Robyn was attempting to form an organization to be a networking center for community supported farms. It is hoped that Robyn and others will be able to travel and make presentations on the CSA concept.

Community Supported Agriculture of North America
c/o Robyn Van En
Indian Line Farm
RR3 Box 85 — Jug End Road
Great Barrington, MA 01230
413-528-4374

INTERNATIONAL ALLIANCE FOR SUSTAINABLE AGRICULTURE is a non-profit organization of individuals and groups cooperating to develop economically viable, ecologically sound, socially just, and humane agricultural systems around the world. It engages in research and documentation, network building, and education.

International Alliance for Sustainable Agriculture
1701 University Ave S.E. Room 202
Minneapolis, MN 55414

The **LAND TRUST ALLIANCE** is a national communications
network and service center dedicated to helping land trusts do the
best job possible in conserving important land resources.

Land Trust Alliance
900 Seventeenth St. NW, Suite 410
Washington, DC 20006-2596
202-785-1410

The **NEW ENGLAND SMALL FARM INSTITUTE** is a small,
non-profit organization dedicated to promoting the sustainable use
of the region's agricultural resources. It manages 400 acres of public
forest and farmland, works to refine viable small-farm management
systems, and to provide information and training in ecologically
responsible commercial production of food and fuel.

New England Small Farm Institute
Box 937
Belchertown, MA 01007
(413) 323-4531

Farms Profiled in this Book

Brookfield Farm
4 Hulst Rd.
Amherst, MA 01002

Codman Farm
c/o Stan White
Lincoln, MA 01773

Greenpower Farm
c/o Weston Youth Commission
Weston Town Hall
Weston, MA 02193

Hawthorne Valley Farm
RD 1 Box 227
Ghent, NY 12075

Indian Line Farm
RR3, Box 85
Great Barrington, MA 01230
Kimberton CSA
c/o Kerry and Barbara Sullivan
P.O. Box 192
Kimberton, PA 19442

Natick Community Farm
117 Eliot St.
South Natick, MA, 01760
508-655-2204)

Sunways Farm — Hugh Ratcliffe
P.O. Box 147
Housatonic, MA 01236

The Temple-Wilton Community Farm
c/o Anthony Graham
RFD 1
Temple, NH 03086

The Watatic Community Farm
RD 2 Box 739
New Ipswich, NH 03071

APPENDIX C
Community Land Trusts

"The land, the earth God gave to man for his home, sustenance and support, should never be the possession of any man, corporation, society, or unfriendly government, any more than the air or water, if as much. An individual or company . . . requiring land should hold no more than is required for their home and sustenance, and never more than they have in actual use in the prudent management of their legitimate business, and this much should not be permitted when it creates an exclusive monopoly. All that is not so used should be held for the free use of every family to make homesteads, and to hold them as long as they are so occupied."
—Abraham Lincoln, Washington, D.C. 1862

This book has clearly presented the concept that the care of the Earth should be the concern and responsibility of everyone. However, current land use is largely determined by those who pay the most for the land, regardless of whether they have the best intentions for it. The purpose of this section is to present several examples of land preservation that recognize that farming and land stewardship is in the best interest of everyone.

Land Owned by Non-Profit Tax-Exempt Organizations for the Purpose of Education, Research, Land Preservation and Farming

Brookfield Farm and the Bio-Dynamic Farm Land Trust

In this case the Land Trust is incorporated as a non-profit tax-exempt [501(c)(3)] organization dedicated to education, research, and preserving farm land. It owns the farm and makes it available to the Brookfield Community Farm. The land is owned by the Trust and is guaranteed to be available for farming in perpetuity. A Board of Directors oversee the activities of the trust and can provide long-term leases to the Community Farm. Since no mortgage remains on the land, it provides access to low and middle income farmers who otherwise would not have the possibility to farm. All gifts to the Trust are tax-deductible.

156

While this set-up is almost ideal, you can't assume that it can be transplanted to other parts of the country. The Internal Revenue Service grants tax-exempt status to organizations created to serve the public good via scientific, educational, religious, and charitable purposes. Research farms or educational farms do fit the code, but the preservation of farmland in itself is not viewed as meeting the criteria. However, in some parts of the country preservation of farmland may meet IRS criteria if the farm lies in an ecologically sensitive zone, is a habitat for endangered wildlife, has special unique scenic value, is specifically set up for low-income families, or provides "open space" in an area of residential development. Brookfield's location near the town drinking water supply helps its case for a favorable tax ruling.

Camphill Village Farms

Camphill Villages are non-profit tax-exempt organizations created for the principal purposes of working with mentally handicapped people and working the land biodynamically. Like all 501(c)(3) organizations they can receive tax-exempt gifts of land, equipment, machinery, animals, and grants. There are currently six Camphills in the U.S. ranging in size from 20-500 acres. In these cases the land is preserved for agriculture. Some farmers stay and work a lifetime while others may stay shorter periods. No equity is acquired by the individuals who work, but the members of the community provide for the needs of retired farmers as long as they reside in the village. In this way the security in old age depends on human beings rather than bank accounts and the Social Security Administration.

Community Land Trusts

Most Community Land Trusts (CLT's) are based in urban areas to acquire housing for low-income families. The houses are owned by the Trust for the public good, and long-term leases are provided to tenants. For example, the Community Land Trust in the Southern Berkshires was created in 1980 under the wise guidance of Robert Swann to bring together low-income housing and modest agricultural use.

In this model the Trust owns the land and provides 99-year leases to families who build and own their homes on the land. All improvements to buildings or land are owned by the individuals who make them. If a family wishes to move, then the house and improvements can be sold but not at the highest market value, which would exclude future low-income people. The selling price of a house is determined by using the original building cost plus improvements and inflation, then subtracting depreciation. In this way no speculative value or other false economic value can affect the price.

The agricultural land is made available in the same way. Generally speaking, CLT's are non-profit corporations but not tax-exempt. This has the disadvantage that gifts are not tax-deductible. However, it has the advantage that the lessee can own improvements and build equity. It also avoids the growing degree of IRS regulations and restrictions on 501(c)(3)'s.

A thorough treatment of CLT's can be found in *"The Community Land Trust Handbook"* by the Institute for Community Economics.

Land Trusts that Protect Farming by Holding Development Rights or Agricultural Easements

The American Farmland Trust (AFT) is perhaps the best known example of this. It is also a 501(c)(3). In many parts of the country, land prices have escalated due to residential and commercial development rather than increased agricultural potential. Farmers, especially young beginning ones, are unable to pay the inflated prices of farmland.

In many cases, retiring farmers would prefer to see the land they love carry on in agriculture rather than in housing developments. By donating an agricultural easement to AFT, the land is preserved for farming in perpetuity. The farmer still owns the land; but he/she has given away the right to develop the land. As a result, the value of the property drops to its agricultural value and when the farm is sold, a farmer can afford to buy it. Since the original donor has given away a real value with the easement, it is considered a tax-deductible donation. The difference between the development value and the agricultural value is the amount of the gift. AFT is also a strong supporter of organic agriculture.

An excellent booklet on the value of agricultural easements to protect farmland and the forced sale of farmland due to estate taxes is *"Preserving Family Lands"* by Stephen Small, P.O. Box 146, W. Peterborough, NH 03468.

Government Purchase of Development Rights (PDR)

Most states in the Northeast recognize that preserving farmland is in the interest of the public good and have set up funding to purchase development rights from farmers. In the model stated above, the development rights are donated by the farmer. With the PDR program the state buys the rights from the farmer because it recognizes that all farmers aren't in a financial position to be able to give them away. Usually the farmers offer their development rights to the state and the state determines which farms it can protect with the funds it has available. AFT has taken a leading role in assisting states with setting up these programs.

Land Conservation Trusts

Land Conservation Trusts (LCT's) are created to protect wildlife areas, open space, and ecologically sensitive areas. They are non-profit tax-exempt organizations that use easements to protect land or own the land and maintain it. In some cases, parts of the property are suitable for farming and these are leased, usually on a short-term lease. The Land Trust Alliance publishes a helpful book titled *"Starting a Land Trust — A Guide to Forming a Land Conservation Organization."*

501(c)(2) — 501(c)(3) Arrangements

One model of land preservation is currently being used by the Ozark Regional Land Trust. Robert Swann of the E.F. Schumacher Society proposed this idea and it has worked well in the Ozarks since 1985. It has been used to blend the positive tax-deductible benefits of LCT's and the equity-building private ownership benefits of Community Land Trusts. The Ozark Regional Land Trust (ORLT) was set up as a 501(c)(3) LCT to protect open space and land stewardship. The IRS allows educational and charitable organizations to create Title Holding Corporations 501(c)(2) to hold title to land and buildings on its behalf. The ORLT has created four separate 501(c)(2)'s to own land and manage it on its behalf. These 501(c)(2)'s are Community Land Trusts and lease the farmland for 99-year renewable and inheritable terms to individuals. The residents can build equity as mentioned in the Section on CLT's above, and ORLT can receive tax-deductible gifts of land, money, or equipment which it can pass over to the CLT's which it owns. ORLT has written a manual on this approach entitled *"New Organizational Prospects for Community and Conservation Land Trust"* by Gregg Galbraith, 427 S. Main, Carthage, MO 64836.

APPENDIX D
Sample Market Rules
1990 Farmer's Market
The Watatic Community Farm

1. The Watatic Community Farm Farmer's Market will operate from 9 am till 12 noon each Saturday in July, August, and September at the Grange Hall on Main St., in New Ipswich.

2. Participation is open to any person interested in its promotion, who abides by the Market Rules.

3. Vendors must arrive at or before 9 am for registration and setup. After checking in, vendors will be assigned a selling space on a first-come, first-served basis by the Market Coordinator.

4. Vendors will be charged 10% of the amount they make from sales each market day. Vendors will be responsible for figuring their own total sales, and calculating the 10%. Payment must be made on market day. The proceeds of this fee will be used to promote and operate the Watatic Community Farm.

5. Each vendor will be responsible for setting up, displaying, and packaging his or her products, as well as protecting those products from the elements. Each vendor must leave his or her selling area in clean and orderly condition. All refuse and unsold goods must be removed from the Market area by the vendor.

6. Small vendors, such as gardeners, may coordinate and sell for each other.

7. Only locally grown or produced food products, flowers, herbs, and baked goods may be offered for sale. Purchased products may not be sold.

8. Baked and processed goods (such as breads, jellies, jams, preserves, pickles) must be prepared by the vendor, and the vendor is responsible for any necessary licenses.

9. Prices should be fair market value, negotiated by the vendor and the customer. The Watatic Community Farm will make a copy of the New Hampshire *Weekly Farmer's Bulletin* available as a general guide for prices. Neither the Watatic Community Farm nor the Watatic Grange is responsible for the arrangements made between vendors and customers. No warranty of any sort, expressed or implied, is made by the Watatic Community Farm.

10. The Watatic Community Farm will assign a Market Coordinator for each week to work with the vendors. The Market Coordinator is the official representative of the Watatic Community Farm. If problems arise, disputes will be settled by the Market Coordinator.

11. No goods are to be sold before the market officially opens.

12. Surplus food may be donated to the Watatic Community Farm, which will make every effort to see that the food gets to a needy group.

13. Any accident or injury must be immediately reported to the Market Coordinator. Anyone who comes to participate in the market, vendor or customer, comes at his or her own risk. Neither the Watatic Community Farm nor the Watatic Grange is liable for an injury to person or property.

14. Violation of the Market Rules may subject a Vendor to exclusion from further participation in the Market following review by the Watatic Community Farm.

15. These rules are intended to be fair and in the best interests of all who participate in the Farmer's Market. The Watatic Community Farm may, at any time, modify or add to these rules to better serve those interests.

APPENDIX E
Sample Budgets*
1987 Estimated Budget for
the Kimberton CSA

MEMBERSHIP

We originally planned for 38 shares at an average cost of $600 (plus two free shares allocated for the apprentices). Two members dropped out, so we ended up with 36 paying shares (39 half shares and 14 full shares).

In the second year (1988) the size of a share was cut in half, since that seemed to be a more popular amount of produce for one household.

Operating Expenses	Projected 1987	Actual 1987
Seed	500	775
Land and bldg. rent	750	750
Tractor rental	200	250
Manure	100	100
Greenhouse fuel	425	298
Tools, supplies, etc.	500	1,750
Salaries	9,500	10,000
House rent	5,500	3,700
House utilities	1,250	802
Medical insurance	2,000	1,685
1/2 FICA	0	1,430
Capital payment	2,000	2,706
TOTALS	22,725	23,562

*NOTE: The members of the Kimberton CSA, as well as all other CSAs, emphasize that the plan and budget for their operation is not directly transferable. Each group must make its own way based on the people and resources available. This budget is presented simply as an example of how one group approached the task with the resources they had at hand.

Income	Projected	Actual
Shares	22,800	22,036
Interest	—	244
Accounts receivable	—	546
Totals	22,800	22,826

DEFICIT...736
(Since we ended the year with a deficit, the shareholders were asked to contribute an additional $19.37 per whole share, or $9.68 per half share.)

Capital Expenses	Projected 1987	Actual 1987
Greenhouse	3,000	1,041.65
Seed flats	400	636.41
Fencing	500	932.04
Irrigation	200	928.60
Tractor	5,000	3,142.90
Bedmaker		850
Weed trimmer		289.48
Seeder		869.48
TOTALS	10,000	8,690.20

STILL NEEDED:
Basket weeder
Disc
Irrigation

Kimberton CSA Operating Budget for 1989

INCOME:

Shares	$37,500
Sales to dairy	1,000
Interest	350

EXPENSES:

Seeds and plants	1,800
Fertilizer	1,000
Supplies (including sprays)	2,200
Repair, maint., tools	2,500
Fuel	500
Equipment rental	350
Land Rent	200
CSA Utilities	470
Office	500
Apprentice Sal. and rent	3,900
Health Insurance	1,700
Salary	14,500
Rent (housing)	2,400
Utilities (residential)	1,000
1/2 FICA	1,250
Capital repayment	4,500
TOTAL	$38,850

CAPITAL BUDGET

Previous balance	3,236.35 (Carried by Sullivans)

1989 expenses

Irrigation	3,000
Truck	2,500
Lath house	200
Greenhouse plastic	300
Fencing	300
TOTAL EXPENSES	$ 6,300 (Carried by members)

THE 1988 AVERAGE HARVEST FOR ONE SHARE
AT THE KIMBERTON CSA*

CROP	PLANNED	ACTUAL
Beans	20 lbs.	18 lbs.
Beets	20 lbs.	29 lbs.
Broccoli	10 lbs.	7 lbs.
Cabbage	30 lbs.	19 lbs.
Carrots	40 lbs.	51 lbs.
Cauliflower	10 lbs.	2½ lbs.
Celery	5 lbs.	4 lbs.
Celeriac	2 lbs.	1 lb.
Chinese Cabbage	12 lbs.	16 lbs.
Corn	60 ears	52 ears
Cucumbers	1 lbs.	18½ lbs.
Eggplant	5 lbs.	5 lbs.
Kale & Collards	5 lbs.	6 lbs.
Leeks	7 lbs.	7 lbs.
Lettuce	50 heads	37 heads
Mustard	3 lbs.	½ lb.
Onions	25 lbs.	15 lbs.
Parsley	2 lbs.	3 lbs.
Parsnips	8 lbs.	8 lbs.
Peas	5 lbs.	2½ + U-pick
Sugarpeas	8 lbs.	10 lbs.
Peppers	15 lbs.	19½ lbs.
Potatoes	30 lbs.	12½ lbs.
Radish	1 lb.	1 lb.
Rutabaga	6 lbs.	0 lb.
Scallions	2 lbs.	½ lb.
Spinach	10 lbs.	19 lbs.
Squash-Summer	15 lbs.	13 lbs.
Squash: Winter + Pumpkins	40 lbs.	31½ lbs.
Swiss Chard	4 lbs.	10 lbs.
Tomatoes	38 lbs.	53 lbs.
Turnips	4 lbs.	6 lbs.
Cantalope	32 lbs.	56 lbs.
Watermelon	25 lbs.	25½ lbs.
Sweet Potato	8 lbs.	none planted

THE 1988 AVERAGE HARVEST FOR ONE SHARE
AT THE KIMBERTON CSA* (Continued)
ADDITIONAL CROPS RECEIVED IN 1988

CROP	PLANNED	ACTUAL
Basil		2½ lbs.
Chives		½ lb.
Dill		½ lb.
Marjoram		½ lb.
Okra		1¼ lbs.
Oregano		⅓ lb.
Daikon Radish		2¼ lbs.
Sage		¼ lb.
Thyme		¼ lb.
Mini Pumpkins		1 lb.

*Note: One share cost $320. The same quantity of vegetables, according to prices at local markets, would have cost $530. Thus, shareholders of the Kimberton CSA saved $210 each in 1988.

About the Authors

Trauger Groh has been a farmer for 33 years and has been at the leading edge of both the biodynamic and community farm movements. After helping to establish a widely known community-supported farm in North Germany, he settled down in Wilton, NH, where he helped to start the Temple-Wilton Community Farm. This farm raises the fruits, vegetables, milk, and eggs for about 85 New Hampshire families. Trauger is also active as a consultant for many other farmers and farming groups in America and abroad. Well known in the international agricultural community, he has lectured hundreds of times on social, educational and agricultural issues.

At present, Trauger is particularly active through the following groups: The Bio-Dynamic Farmers of the North East and the Bio-Dynamic Farming and Gardening Association. Locally, he is active through the Cadmus Corporation, a non-profit organization dedicated to acquiring land and the other resources necessary to enable farmers to farm, and for training and research. Cadmus is dedicated to finding practical ways to help develop the farms of tomorrow. Trauger is actively raising funds to support this endeavor. You may contact him through the Cadmus Corporation, P.O. Box 333, Wilton, NH 03086.

Steven S.H. McFadden lives about five miles down the road from Trauger. A freelance journalist since 1975, Steven has published hundreds of magazine and newspaper articles, and writes a weekly column entitled *Organic Outlook*. One of the founders of the Watatic Community Farm, he is the author of a book entitled *The Legend of the Rainbow Warriors: A Journalist's Account of An Emerging Myth*, and is also writing a book of interviews with Native American elders. The book will be entitled *Profiles in Wisdom: Native Elders Speak About the Earth*.

In addition to agriculture and social progress, he has a longstanding interest in astrology. He is the founder of *Chiron Communications*, a company dedicated to communicating not only the vision of a world that works for everyone, but also the tools to make it real. Through the company, Steven creates and publishes media, presents lectures and workshops, and offers consultations for individuals and organizations. You may contact him through Chiron Communications, P.O. Box 328, New Ipswich, NH 03071.